ST(P) Technology Today
A Series for Technicians

Ma
Technology for TEC Level II

Also from Stanley Thornes (Publishers):

Engineering Instrumentation and Control by D C Ramsay

Engineering Science Level II by G D Redford, D T Rees and A Greer

Physics for TEC Level II by T Lowe, K Roe, G D Redford, D T Rees and A Greer

Engineering Design for TEC Level III by M D Brooks and D Oldham

Mechanical Science – TEC Level III by G W Taylor

A full list of Technology Today titles is available on request

ST(P) Technology Today
A Series for Technicians

Materials Technology for TEC Level II

P F Kiddle
AInstP
Gloucestershire College of Arts and Technology

D C Ramsay
BA(Hons) MIMechE
Gloucestershire College of Arts and Technology

Stanley Thornes (Publishers) Ltd

First published in 1983 by:
Stanley Thornes (Publishers) Ltd
Educa House
Old Station Drive
Leckhampton Road
CHELTENHAM GL53 0DN
England

British Library Cataloguing in Publication Data

Kiddle, P.F.
Materials technology for TEC level II.
1. Materials science
I. Title II. Ramsay, D.C.
620.1'1 TA403

ISBN 0-85950-372-0

Typeset by Tech-Set, Unit 3, Brewery Lane, Felling, Tyne & Wear
Printed and bound in Great Britain by
Ebenezer Baylis & Son Limited
The Trinity Press, Worcester and London

CONTENTS

PREFACE

This book is a work of collaboration between a metallurgist (PFK) and a design engineer (DCR).

It is written primarily to cover the objectives of the TEC unit Materials Technology II (TEC U80/738), but we feel that the unit has been so well designed as an introduction to metallurgy and polymer material structures that the book will also be extremely valuable to any engineer seeking a basic understanding of these subjects.

The first seven chapters of the book correspond to the seven sections into which the TEC unit is divided. The eighth chapter has been added to explain how metallurgical specimens are prepared for examination by hand lens and microscope, and to illustrate the apparatus needed for this purpose.

Mechanical properties which are referred to in the text are explained at the beginning of the book, and the names, symbols and atomic characteristics of the chemical elements are listed in appendices at the end.

The main part of each chapter has been compiled by the metallurgist, but each chapter is followed by a summary contributed by the designer, which sets out again the essential facts that the student should remember. We feel that this approach, illuminating each topic from two slightly different viewpoints, will give the student a clearer insight into the subject than could be obtained from a single statement of the principles involved.

We have illustrated the text wherever possible with diagrams, graphs, photomicrographs and tables. At the end we have provided some crosswords for a little light-hearted revision. We hope that students will find the book as interesting to read as we have found it to write.

P F Kiddle
D C Ramsay
Gloucester 1983.

Introduction: DEFINITIONS OF MECHANICAL PROPERTIES

The tensile test is used primarily to indicate (i) the tensile strength, (ii) the ductility, (iii) the elastic range of the material.

The *tensile strength* is defined as the maximum force sustained by the material before fracture, divided by the original area of cross-section. Thus,

$$\text{Tensile strength (N/mm}^2\text{)} = \frac{\text{Maximum force (N)}}{\text{Area of cross-section (mm}^2\text{)}}$$

Alternatively the force may be recorded in MN and the area of cross-section in m^2 giving the tensile strength in MN/m^2 ($1\ MN = 10^6\ N$). The numerical value is the same in N/mm^2 as in MN/m^2; for example, the tensile strength of a material may be recorded as $600\ N/mm^2$ or $600\ MN/m^2$.

The *percentage elongation value* is used as an approximate measure of the ductility of a material and is defined as the extension of the gauge length after fracture, stated as a percentage of the gauge length. Thus,

$$\text{Percentage elongation} = \frac{\text{Extension after fracture}}{\text{Gauge length}} \times 100\%$$

The value obtained depends on the geometry of the test specimen and the gauge length over which the extension is measured. In the UK the gauge length used is $5.65 \times \sqrt{\text{area of cross-section}}$ or $5 \times$ diameter in the case of a round specimen.

In carrying out a tensile test it is usual to plot a stress–strain curve by noting the extension of the gauge length at a number of force increments. Since

$$\text{Stress} = \frac{\text{Force}}{\text{Original area of cross-section}}$$

and

$$\text{Strain} = \frac{\text{Extension}}{\text{Gauge length}}$$

the corresponding stress and strain values may be determined.

For a few materials, notably very mild steel, the curve may be as shown in Fig. I.1, but the more general case is shown in Fig. I.2. The *elastic limit* of the material is the highest stress which may be applied to the material without permanent deformation taking place. For most practical purposes we may assume that the elastic limit corresponds to the *limit of proportionality*, which is the stress at which the strain ceases to be proportional to the stress, i.e. the stress at which the curve departs from the straight line. Beyond this limit of proportionality the material is said to be *plastic*, and for a stress in this plastic range, removal of the stress will result in some permanent (plastic) deformation. In practice the limit of proportionality value obtained is quite significantly dependent upon the sensitivity of the extensometer used and the draughtsmanship used in 'fairing' the curve to the straight line part of the stress–strain curve. For this reason a *proof stress* criterion is usually used, e.g. the 0.1% proof stress, which is the stress at which the curve departs from the line of proportionality by 0.1% strain (equivalent to an actual strain of 0.001). It is also the stress at which the permanent (plastic) strain of the material is 0.001. In some cases 0.2% proof stress or 0.5% proof stress values are used. In terms of engineering design the 0.1% proof stress is a more realistic design value than the tensile strength of the material. The ratio $\frac{\text{0.1\% proof stress}}{\text{Tensile strength}}$ is significant since it gives an indication of the extent of the plastic deformation which will occur before failure takes place.

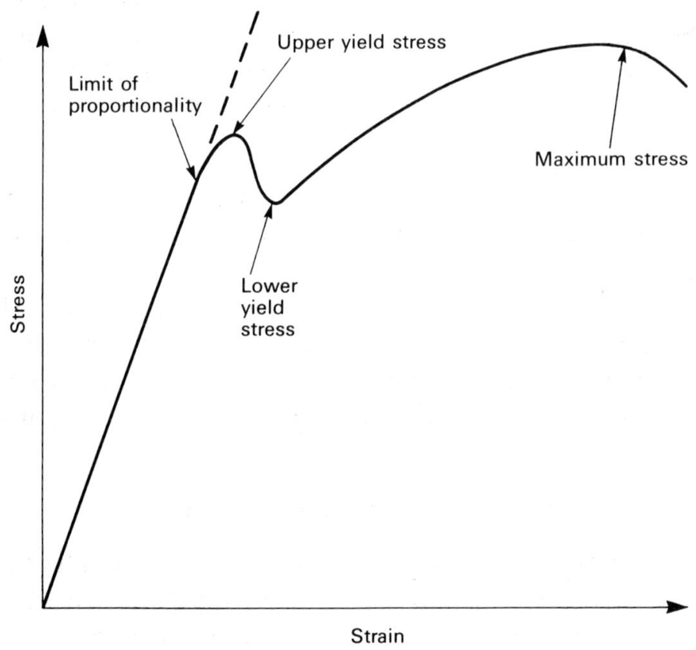

Fig. I.1 Stress–strain curve for very mild steel

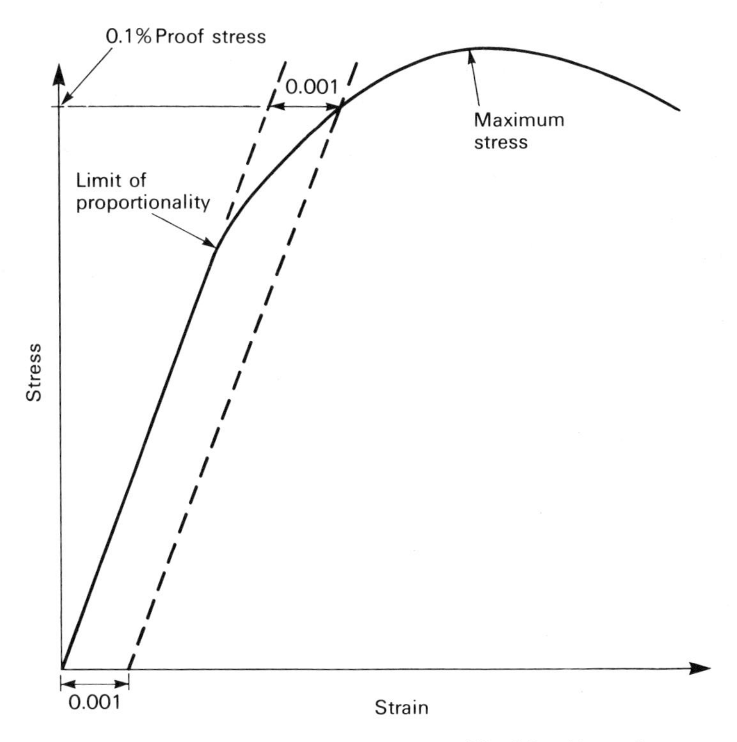

Fig. I.2 Normal stress–strain curve

The slope of the line of proportionality provides a value for *Young's modulus*, which is a measure of the stiffness of the material. Thus, for example, since Young's modulus for steel is approximately three times that for aluminium, then for a given stress the aluminium will deflect three times as much as the steel specimen.

The *hardness* of a material is usually defined in terms of the resistance of the material to indentation using a standard diamond or ball indenter pressed into the material by a given load for a set time. The standard tests are the *Brinell test* (ball indenter) and the *Vickers test* (diamond indenter). While the hardness value will give only an approximate measure of the wear resistance of the material, it is also a valuable non-destructive quality control tool, used to assess the tensile strength of a material. For most metals it has been shown that

$$\text{Hardness value} \times \text{Factor} = \text{Tensile strength}$$

where the factor depends on the nature of the metal and its heat treatment condition. Since for most applications the indentation made by the hardness test is not significant, the test provides a means of assessing the tensile strength of finished components.

The *Izod test* is an impact test carried out on a special notched test specimen, to determine the toughness or brittleness of a material. The test may also be used to obtain confirmation that heat treatment procedures have been correctly carried out.

The test is carried out by an Izod machine in which a hammer falls through an arc to strike the end of the specimen. The Izod number of the material is the energy, in joules, absorbed by the specimen as it bends or breaks. A low value indicates brittleness; a high value indicates toughness.

Note that in tables of properties, the symbols < meaning 'less than' and > meaning 'greater than' have been used throughout the book.

1 THE CRYSTALLINE STRUCTURE OF METALS

ATOMS

If we could divide and subdivide any piece of any material into smaller and smaller portions, continuing the process indefinitely, we would eventually get down to something that could not be divided any further (at least, not in the sense of cutting it in half): we would be down to an *atom*.

We could not divide it any further because it would turn out to be a system of particles travelling in circular orbits around a central nucleus, rather like a tiny solar system – impossible to divide without turning it into something completely different.

This, then, is an atom: particles of mass with negative charge (*electrons*) in orbit around a larger particle with positive charge (the *nucleus*). An atom in its normal state has just enough electrons to exactly neutralise, with their negative charge, the positive charge of the nucleus. Such an atom is thus electrically neutral.

The electrons orbit the nucleus at radii which are determined by the laws of atomic physics. Since an electron at any instant may be anywhere on an imaginary spherical surface concentric with the nucleus, electrons are said to be in spherical *shells* of various radii. However, there is nothing definite about the position of an electron. The 'shells' represent the *most probable* radii at which electrons may be found at any instant. This idea is illustrated in Fig. 1.1, in which the density of dots at any particular radius represents the probability of an electron's being at that radius at any particular instant.

Fig. 1.1 is a representation of something which no one has ever seen or is ever likely to see, because an atom consists almost entirely of empty space, and the diameter of its outer 'shell' is only about one ten-millionth of a millimetre (10^{-7} mm), whereas the wavelength of the light by which we see is about five thousand times as large as this. The idea of an atom which we now have is the result of remarkable achievements of deduction by scientists over the last hundred years. The nearest we can get to 'seeing' is to use

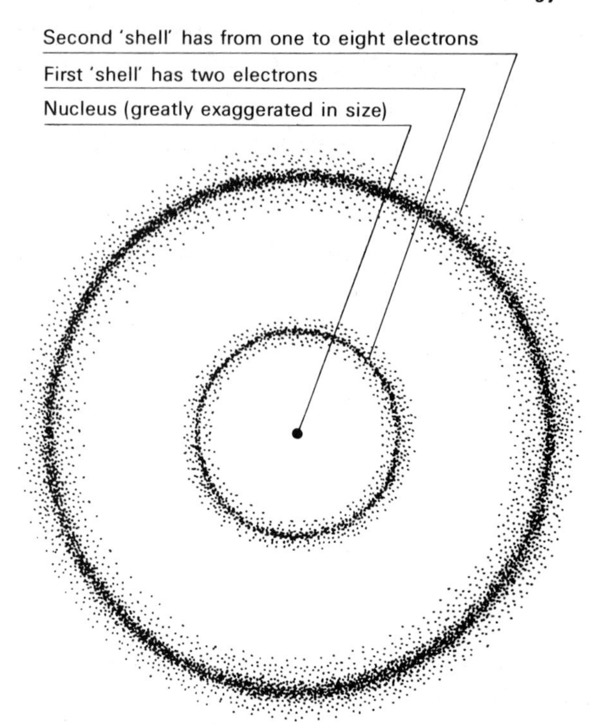

Fig. 1.1 Section through an atom with two 'shells'. The density of dots at any radius represents the probability of an electron's being at that radius at any instant

X-rays, with wavelengths only a few times larger than the diameter of an atom, to deduce the arrangement of atoms in the space lattices of crystals, by X-ray diffraction.

SOME 'ELEMENT'-ARY CHEMISTRY

The millions of different substances, natural or synthetic, in the modern world are made up of the atoms of less than a hundred different *elements*. These are listed in Appendix 1 on p. 161. Each element's distinctive chemical and physical properties are determined by the number of electrons it normally has in orbit around the nucleus of its atom, and by the corresponding mass and positive charge of its nucleus.

Except for a few instances such as copper or carbon, which are pure or virtually pure elements, all of the substances we normally meet with are chemical compounds of two or more elements, or else are mixtures of elements and/or chemical compounds.

In a *chemical compound* two or more atoms bond together in fixed arrangements, called *molecules*. A molecule is the smallest quantity of a compound which can exist on its own. Its atoms may be of the same element – for example, oxygen (O_2), in which two atoms of oxygen bond to each other to form an oxygen molecule. But more usually the atoms are of different elements – for example, in a molecule of water, an atom of oxygen bonds to two atoms of hydrogen. The resulting molecule has the chemical formula H_2O – thus the chemical formula of a compound tells us how many atoms of each element are combined in the molecule of the compound.

A compound has properties which are quite different from those of the elements which went to make it up. It thus differs from a *mixture*: in a mixture the constituents may be in any proportions, and they retain their individual properties after mixing.

HOW ATOMS BOND TOGETHER

All atoms and molecules are attracted to each other by forces much stronger than gravitational attraction, if they get to within a few atom diameters of each other. These forces are *electrostatic* forces, positive and negative charges attracting each other. Electrostatic forces seem feeble when we see them demonstrated in the laboratory, but at a separation of only one atom diameter, and multiplied by the millions of millions of atom bonds which are present in a typical cross-section of material, they can add up to immense tensile strength.

Atoms and molecules can bond together electrostatically in four different ways. The particular way in which the atoms of two given elements bond together depends on whether the outer 'shell' of each atom has a full quota of electrons, or nearly full, or nearly empty, and whether the outer shell electrons are only loosely held (as they are in a metal).*

Thus the atoms of any given elements, or of the same element, will bond together in one particular way, and that particular arrangement of atoms, repeated over and over again, causes a repeated three-dimensional geometrical pattern of atoms to build up. A solid substance in which the atoms are arranged in a regular pattern in this way is called a *crystal*.

*Appendix 2 on p. 163 lists the elements in ascending order of number of electrons per atom, and also gives the number of electrons in each 'shell' of the various atoms. From a study of the shell populations of the first few lines of elements in Appendix 2, it can be seen that for each shell of an atom there is a limit to the number of electrons it can hold.

UNIT CELLS AND SPACE LATTICES

Some substances are obviously crystalline. For example, granulated sugar, viewed through a hand lens, can be seen to consist of clear rectangular blocks, which are the repetitions, on a millionfold scale, of the basic arrangement of atoms in the sugar molecules. Metals, also, crystallise when they solidify; but because their crystals interlock in a single solid mass, their crystalline nature can only be seen when a polished and etched cross-section is examined under a microscope (see Fig. 1.2).

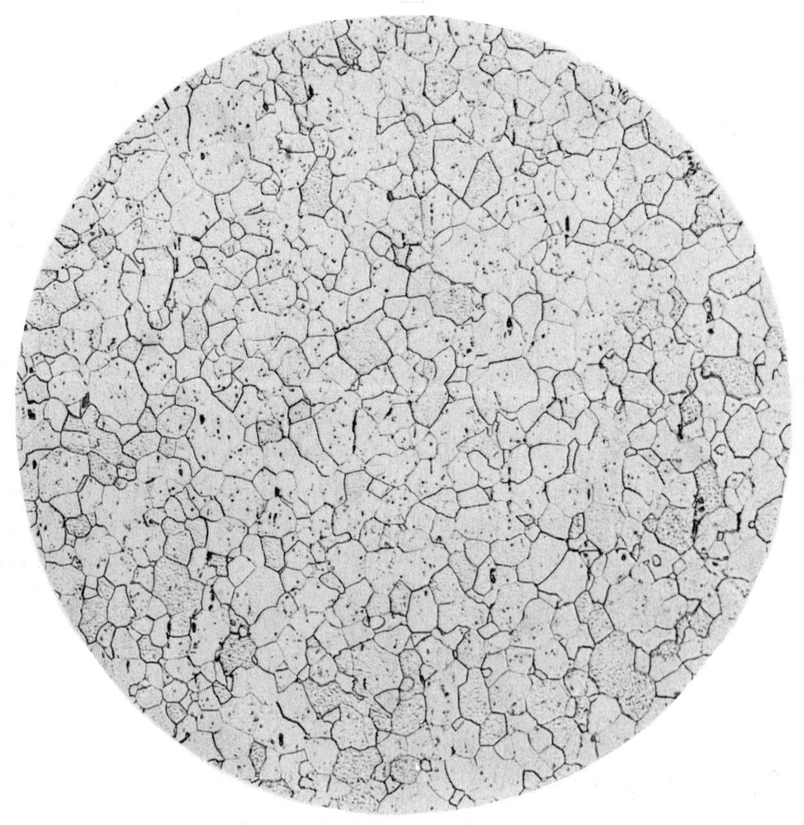

Fig. 1.2 Interlocking crystals in a metal. The photograph shows the grains of ferrite (see Chapter 3) making up a low-carbon mild steel (magnification × 200)

If we could see the actual atoms in such a cross-section, we should see something like Fig. 1.3. We should find that the atoms were arranged in a regular pattern in any one crystal, but that the directions along which they were arranged would be quite haphazard in adjacent crystals. Also the boundaries between the crystals would be quite erratic, because the boundaries occur where the crystals have had to stop growing because there was no room left between them. (The growth of crystals is described more fully on pp. 12–14.)

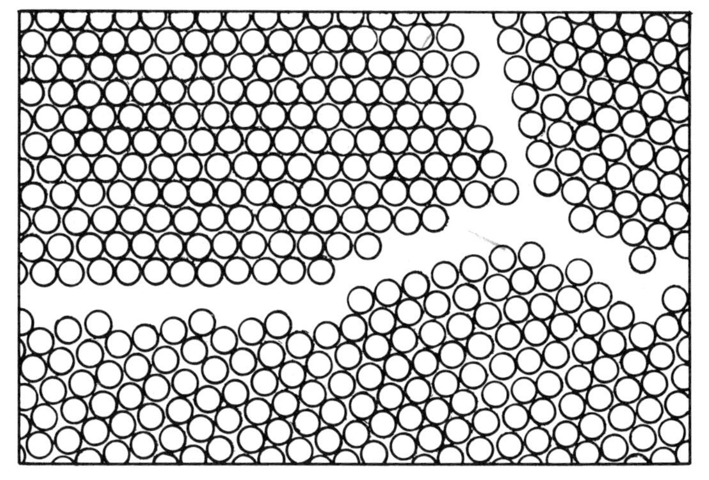

Fig. 1.3 Diagram illustrating the principles of crystal formation in a metal, showing: (i) planes of atoms in random directions in three adjacent crystals; (ii) irregular boundaries between crystals due to crystal growth having to stop where crystals meet (grain boundaries)

The basic three-dimensional arrangement of atoms which repeats itself indefinitely in a crystal is called the *unit cell*, and the structure of unit cells which makes up the crystal is called the *space lattice*. Fig. 1.4 shows a typical unit cell of a crystal with the atoms represented as spheres just touching one another. Presented like that, it probably looks absolutely shapeless. But if we connect the centres of the spheres with straight lines, as in Fig. 1.5, we can see that their basic shape is a cube. And when we

Fig. 1.4 Cluster of atoms forming a unit cell

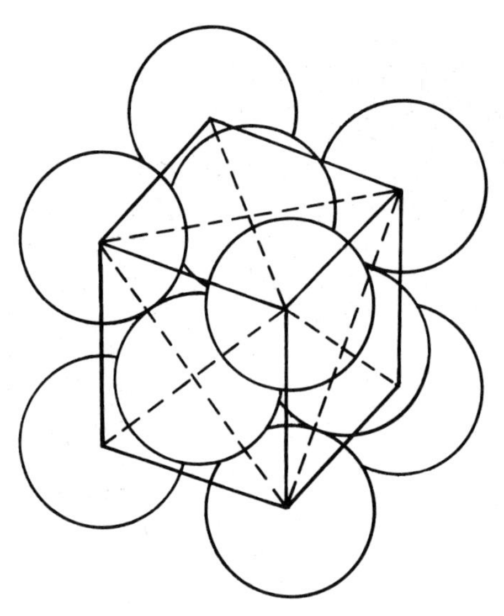

Fig. 1.5 Basic arrangement of a unit cell picked out by dashed lines

remember that the atoms are not spheres at all, but consist mainly of volumes of empty space defined by the paths of electrons, it makes more sense to put a dot at the centre of each atom to represent the nucleus and just join the dots with straight lines to represent the arrangement of the atoms in the unit cell, as in Fig. 1.6. The space lattice of the crystal can then be represented as in Fig. 1.7, by building identical unit cells on to the original one, in all directions, as shown.

The example we have used to illustrate the term 'unit cell' is called a *face-centred cube* (f.c.c.), because it is in the form of a cube defined by the centre of an atom at each corner, with an extra atom at the centre of each face. We have to try to visualise the space lattice as consisting of planes of atoms stretching unbroken along three mutually perpendicular axes, and also along three other mutually perpendicular axes at 45° between the first

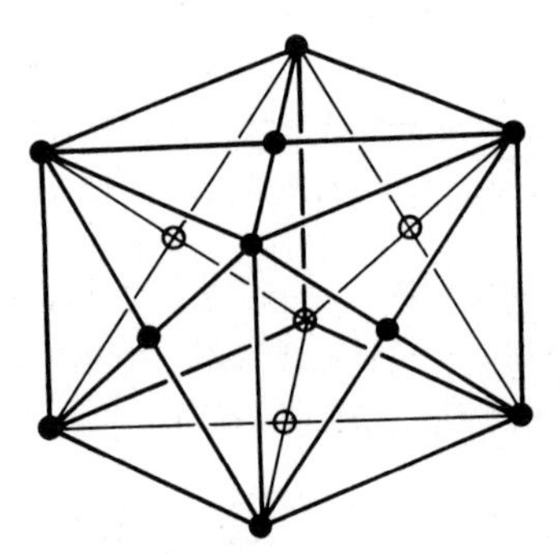

Fig. 1.6 Diagram of the arrangement of the unit cell (this particular example being a face-centred cube)

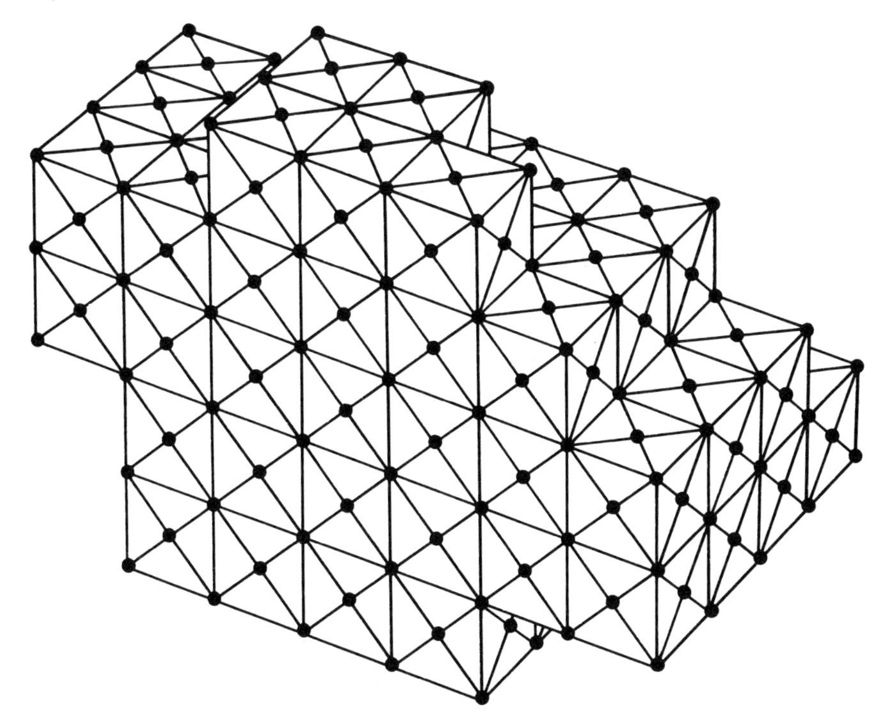

Fig. 1.7 Space lattice of face-centred cubes

three. Then every atom in a face-centred cube type of crystal is bonded to twelve neighbouring atoms, each of which is one atom diameter away, in opposite directions along each of the above six axes. The bonding is the result of the sharing of electrons between the outer shells of the atoms.

Another way in which each atom can be bonded to twelve neighbours gives us the *closed-packed hexagonal* unit cell. We can think of the space lattice of this type of crystal as being made up of parallel layers of atoms arranged in the form of interlocking hexagons, each with an atom at the centre of the hexagon, as in Fig. 1.8. The atoms of the next layer above and below are centred on the spaces between these atoms, as indicated by the dashed hexagon in Fig. 1.8. The atoms in the next layers, above and below that again, revert to the original positions. And so on.

If we consider the relationship between the atoms *a*, *b*, *g*, and *h* in a view looking along the layers, we can see that the lines joining their centres form a tetrahedron (i.e. a three-sided pyramid) as in Fig. 1.9.

Thus the unit cell of a close-packed hexagon type of crystal is made up of the atoms *a*, *b*, *c*, *d*, *e*, *f* and *g* of alternate layers, together with atoms *h*, *m*, and *n* of the layer between them, as shown in Fig. 1.10.

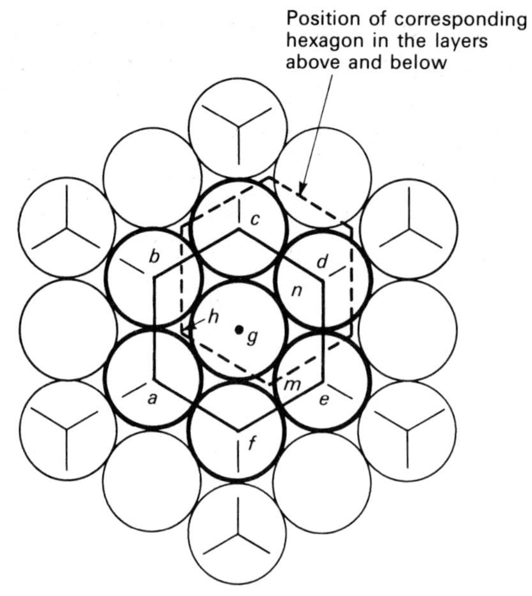

Fig. 1.8 Layer of atoms forming part of a close-packed hexagonal unit cell

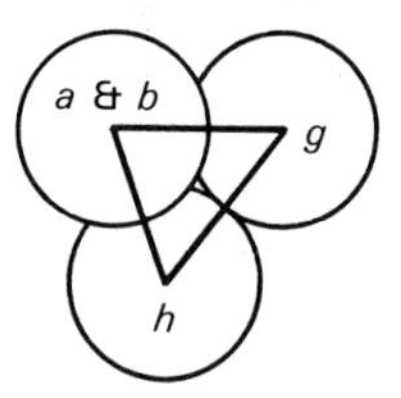

Fig. 1.9 View of atoms *a*, *b*, *g* and *h*, looking along the layers

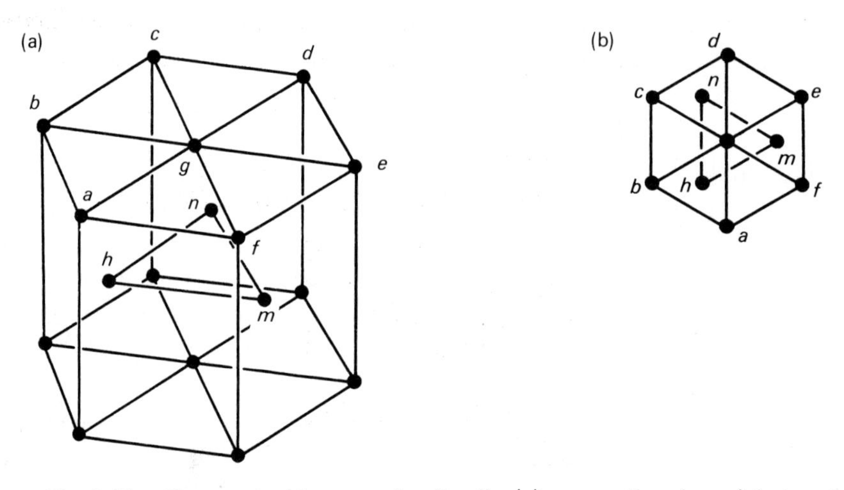

Fig. 1.10 Close-packed hexagonal unit cell: (a) perspective view; (b) plan view

Another type of unit cell is that formed by atoms which require to be bonded to only eight neighbouring atoms. Each atom is then at the centre of a cube. The corners of the cube are the centres of the eight atoms which surround the central atom. This gives us the *body-centred* cubic unit cell (b.c.c.), as in Fig. 1.12. Each of the atoms at the corners of the cube is at the centre of another cube of eight atoms, and so on. The distance between the centres of the central atom and the corner atoms is one atom diameter. The atoms at the corners are father apart from each other than they are from the centre atom, so they do not bond to each other.

Examples of metals which have unit cells in the form of body-centred cubes are chromium, iron (at room temperature) and tungsten. Metals which have face-centred cubes as their unit cells include aluminium, copper and nickel. Metals which have close-packed hexagonal unit cells include cadmium, magnesium and zinc.

The *simple cubic* unit cell shown in Fig. 1.11 is very rare and is probably only exhibited by the element polonium at certain temperatures.

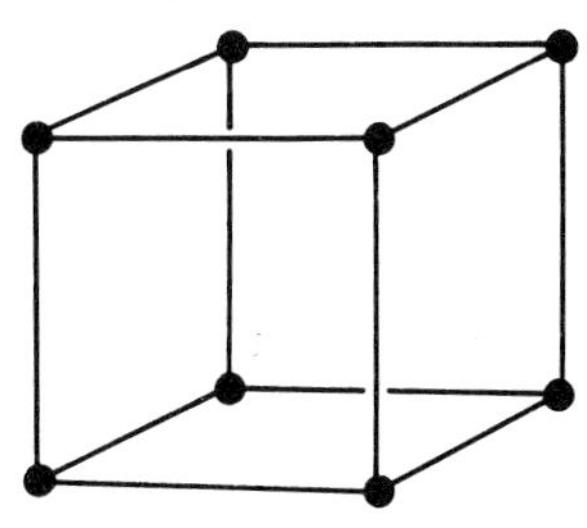

Fig. 1.11 Simple cubic unit cell

Some metals exhibit different unit cell forms at different temperatures, the most important, metallurgically, being iron, which has the body-centred cubic cell at temperatures up to 910°C, face-centred cubic cell between 910°C and 1400°C and reverts to body-centred cubic at 1400°C, i.e.

below 910°C: b.c.c., alpha-iron (αFe)
910°C to 1400°C: f.c.c., gamma-iron (γFe)
above 1400°C: b.c.c., delta-iron (δFe)

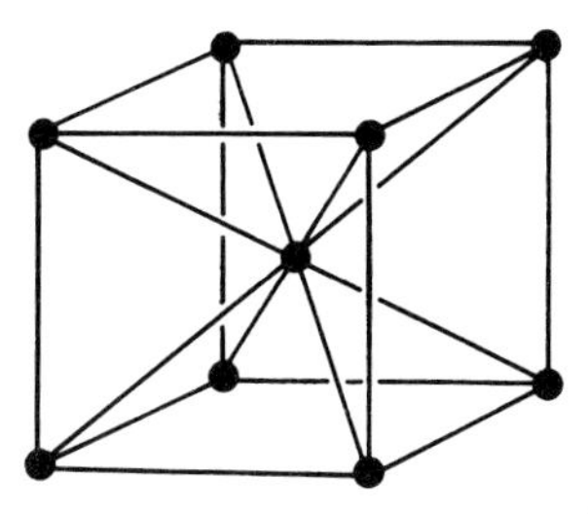

Fig. 1.12 Body-centred cubic unit cell

The face-centred cubic unit cell represents a closer packing of atoms than is the case for the body-centred cubic unit cell and the changes for iron (known as *allotropic modifications*) do in fact result in changes in volume which may easily be observed in the laboratory. If we take a cylinder of iron and measure its volume at various temperatures, then a curve of the type shown in Fig. 1.13 is obtained. (The volume change may be determined from length measurements, because volume and length changes are proportional.) The significant features of this curve are that at 910 °C there is an abrupt decrease in volume associated with the change from b.c.c. to f.c.c., and at 1400 °C there is an equally abrupt increase in volume associated with the change from f.c.c. to b.c.c.

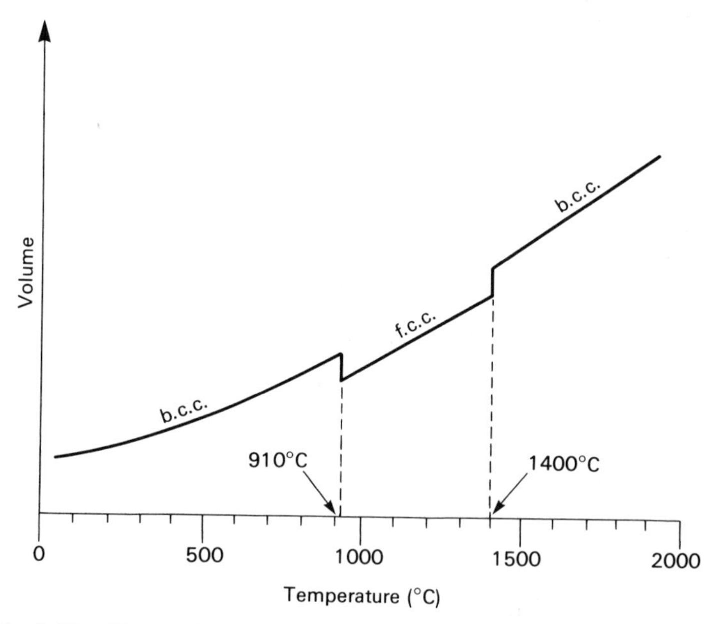

Fig. 1.13 Changes in volume as the unit cell of iron changes from body-centred to face-centred and back again

The type of unit cell associated with a metal is an important factor in determining the mechanical properties of the metal. For example:

face-centred cubic metals are ductile;

body-centred cubic metals have intermediate ductility;

close-packed hexagonal metals are brittle.

THE CRYSTALLISATION OF A PURE METAL

When a pure molten metal cools, it solidifies at a fixed temperature, the melting point or freezing point of the metal. Once the process of solidification starts, the temperature of the metal does not fall any further until it is complete, even though the metal is still giving up heat energy to its surroundings. This is because, in going from the liquid to the solid state at the same temperature, the atoms have to lose a considerable part of their kinetic energy of vibration. This energy given up at constant temperature is known as the *latent heat of solidification.**

If we therefore plot a graph of temperature against time while the liquid metal cools and solidifies, it will have the shape shown in Fig. 1.14. However, the graph will only have this shape:

(i) if a few particles of some other material are distributed through the liquid; and
(ii) if the cooling is fairly slow.

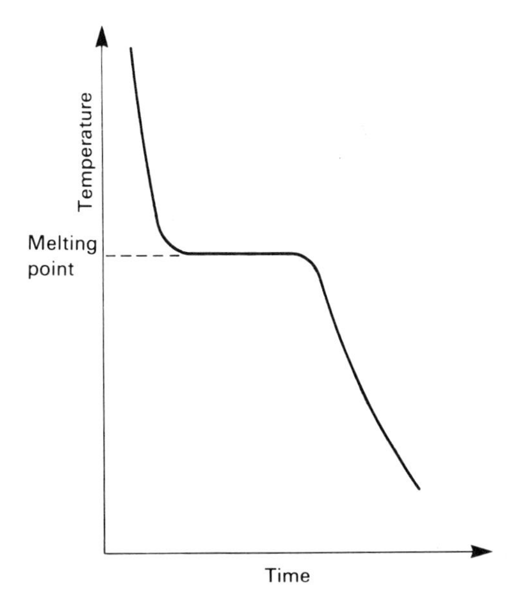

Fig. 1.14 Solidification of a pure metal when solid foreign particles are present to initiate nucleation

If the liquid metal is absolutely pure, so that it contains no 'foreign' particles, the first atoms of the unit cells have nothing to attach themselves to and the liquid may cool to a temperature well below its freezing point before it starts to solidify. When solidification starts, the temperature can rise again to the true freezing point while the rest of the metal sets. This behaviour, which is called *nucleation undercooling*, is illustrated in Fig. 1.15.

*Or *latent heat of fusion*, if we think of melting the metal instead of solidifying it – in that case, the same amount of heat energy is *absorbed* at constant temperature.

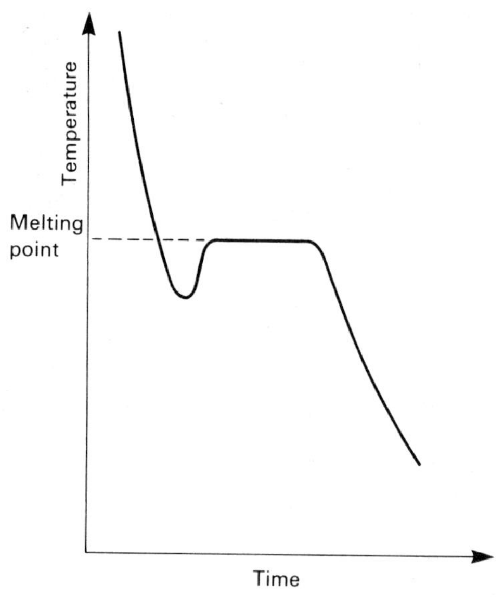

Fig. 1.15 Solidification of a pure metal, showing undercooling due to delayed nucleation

Most commercial supplies of 'pure' metal contain sufficient impurities to prevent nucleation undercooling. Where this is not the case, the metal may be 'seeded' with solid elements or compounds which have a similar crystalline structure to that of the metal, to start off the process of forming unit cells in the liquid. These added substances are sometimes called *grain-refining agents*, because, by promoting the formation of crystals at many points in the liquid, they cause the final solid metal to contain many more crystals of much smaller size; i.e., they give the solid metal a finer grain structure.

When a molten metal is cooled quickly, as in diecasting, solidification takes place at a temperature below the normal freezing point of the metal, even when there are sufficient impurities present to prevent nucleation undercooling. This effect is called *solidification undercooling*; it also tends to give finer grain structures.

What is a 'grain structure'? Suddenly we seem to be writing about wood instead of metal. In fact, the analogy is not so far-fetched. To understand the similarity, we need to consider what happens when molten metal sets. Fig. 1.16 illustrates the process: (a)–(e) represent successive stages in the solidification of a 'pure' metal, as would be seen at a horizontal plane through the metal.

Fig. 1.16(a) represents the formation of the first unit cells in the molten liquid at centres of nucleation which, as we have seen above, may be artificially induced by particles of impurities.

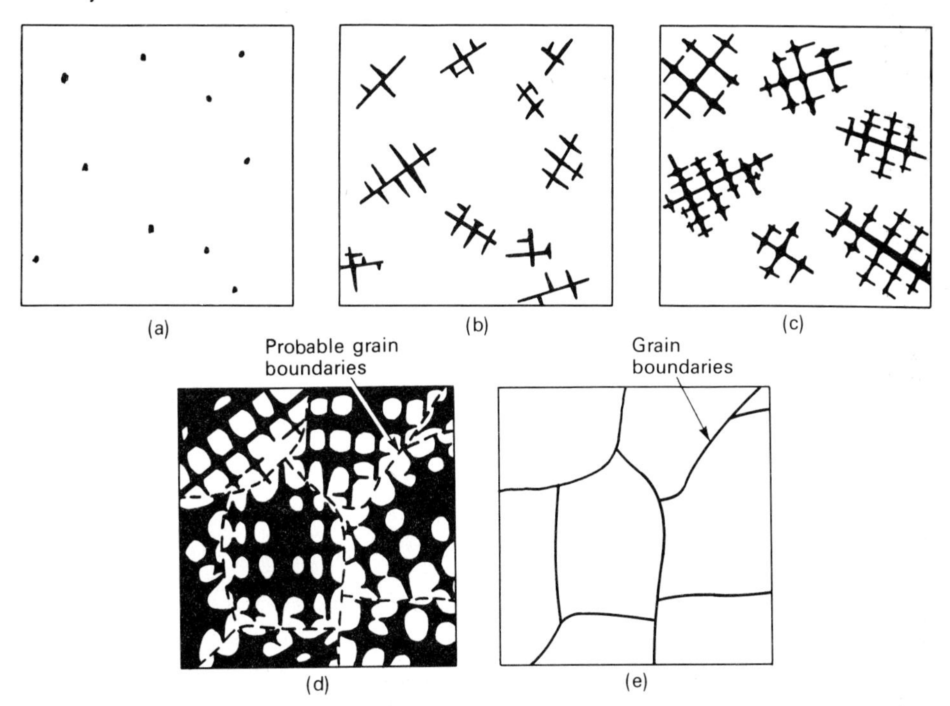

Fig. 1.16 Stages in the solidification of a metal: (a), (b) and (c) are the initial stages of solidification from nuclei; (d) shows the completion of solidification and formation of grains

Fig. 1.16(b) shows the growth of small space lattice structures made up of unit cells built on to one another, but the building up does not occur uniformly all round the original unit cell. The original cells tend to build out two spikes on opposite sides, and then suddenly start building out at right angles again, at points along the original spikes, while continuing to grow in the original directions. These new projections also start growing out at right angles at intervals, and so on. So spiky solids, called *dendrites*, of the form shown in Fig. 1.17, start to form in the liquid metal.

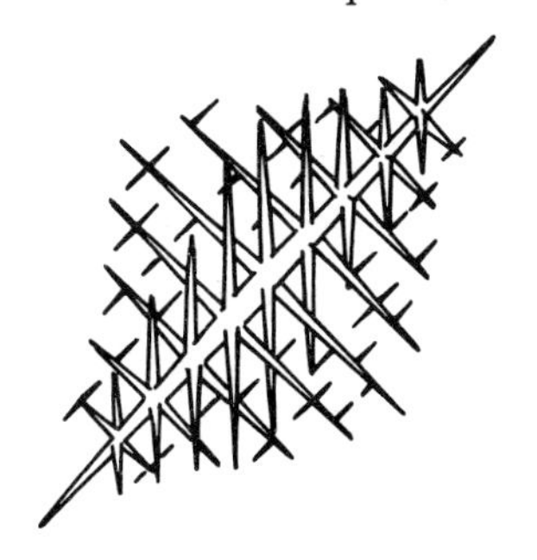

Fig. 1.17 Dendrite

As the cooling progresses, the dendrites grow, and begin to fill in the spaces between their innermost branches, as in Fig. 1.16(c). Eventually they begin to encroach on each other's 'territory'; at this point the final boundaries of the crystals can be deduced, as in Fig. 1.16(d).

Finally the last of the liquid sets, and the metal consists of a solid mass of interlocking, irregularly shaped crystals – this is the *grain structure* of the metal, as in Fig. 1.16(e).

The boundaries between the crystals (the *grain boundaries*) show up because insoluble impurities tend to be precipitated outside the crystals.

Fig. 1.18 is a photograph of a cross-section of aluminium–silicon alloy in which an aluminium-rich solution, solidifying first, formed dendrites in the still-molten remainder of the alloy.

Fig. 1.18 Modified aluminium–silicon alloy structure showing aluminium–rich solid–solution dendrites (white) in unresolved eutectic matrix (magnification × 200)

The dendrites (and hence the planes of the space lattices) lie in random directions, so the grain structure of a metal finally has space lattices running in all directions. This gives the metal uniform strength in all directions – if a piece of metal consisted of a single large crystal, it would be particularly weak in directions related to the planes of the space lattice. For this reason it is desirable for metals to have as fine a grain structure as possible.

SUMMARY

- An *atom* is the smallest possible quantity of any element.
- Atoms consist of a nucleus (positive) orbited by electrons (negative), in various 'shells'.
- A *molecule* is the smallest possible quantity of any compound.
- Atoms and molecules bond together to form solids in ways which depend on the relative fullness of each atom's outer shell.
- Many solids, especially metals, are crystalline – i.e., they solidify into an orderly arrangement of atoms or molecules which repeats continuously throughout the crystal.
- The basic arrangement of atoms which repeats in a crystal is called the *unit cell*. The large-scale orderly arrangement formed by these repetitions of the unit cell is called the *space lattice*.
- The three main kinds of unit cell formed by metals are summarised in Fig. 1.19.

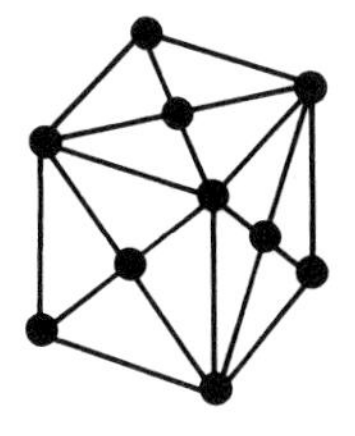

Face-centred cube
Ductile metals.

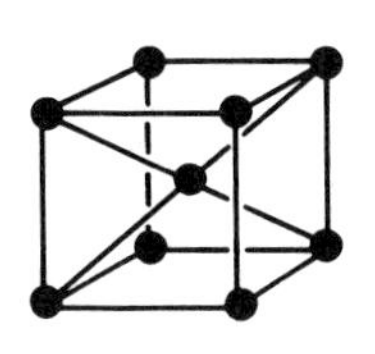

Body-centred cube
Metals of intermediate ductility.

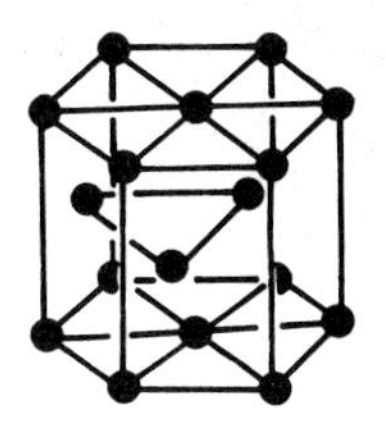

Close-packed hexagonal
Brittle metals.

Fig. 1.19 Three kinds of unit cell

- When a molten metal cools it starts to solidify by forming tiny crystals at *centres of nucleation* throughout the liquid. These crystals grow as *dendrites*, spiky solids which branch at right angles. Eventually the dendrites grow big enough to meet, then the last of the molten metal sets between their branches. Each dendrite has become a crystal, and the crystals interlock to form the solid metal. This arrangement of interlocking crystals is called the *grain structure* of the metal. Between adjacent crystals is the *grain boundary*, a very thin separation surface which tends to contain insoluble impurities. For uniform strength throughout a metal, its grain structure should be as fine as possible.

- When a molten substance starts to solidify, its temperature remains constant until solidification is complete. This is because it is giving out the latent heat which it absorbed at constant temperature when melting – the *latent heat of fusion*.

EXERCISE 1

✓1) (a) Describe, with the aid of a sketch, the arrangement of the components of an atom.

(b) How are the atoms of the various chemical elements distinguished from one another?

(c) What determines the way in which the atoms of elements hold together to form solid material or chemical compounds?

(d) (i) Explain what is meant by the term *shell* as it applies to an atom. (ii) What is the largest number of shells in an atom? (ii) If the first two shells of an atom are full, what are their contents?

2) What is the essential feature of the grouping of the atoms in: (a) a polymer; (b) a metal?

3) (a) What is the essential feature of the arrangement of atoms in a crystal?

(b) Explain, with the aid of sketches, what you understand by the terms: (i) unit cell; (ii) space lattice.

✓ 4) (a) Sketch the arrangement of atoms in the following unit cells: (i) body-centred cubic; (ii) face-centred cubic; (iii) close-packed hexagonal.

(b) Explain, with the aid of a diagram, how the atoms are arranged, in successive layers, when close-packed hexagonal unit cells are formed.

5) (a) What are the changes which occur in the unit cell of iron, as the temperature of iron is varied from room temperature to its melting point? Relate these changes to a sketched graph of temperature against volume.

(b) What is the general name for such changes?

(c) How may the changes be demonstrated in the laboratory?

6) (a) What types of unit cells are associated with (i) ductility, (ii) partial ductility, (iii) brittleness in a metal?

(b) Explain what implication this has for the range of temperatures within which iron is most easily shaped by forging.

7) Explain what the following terms mean:

(a) latent heat of solidification; (b) nucleation undercooling; (c) grain refining agents; (d) solidification undercooling. Illustrate your answers to (a) and (b) with sketches of the appropriate cooling curves.

8) (a) Explain, with the aid of a sketch, what a dendrite is.

(b) Explain, with sketches, the successive stages in the solidification of a pure metal.

(c) What do you understand by: (i) the grain structure of a metal; (ii) grain boundaries?

9) Explain why it is desirable for metal to have as fine a grain structure as possible.

2 EQUILIBRIUM DIAGRAMS

MIXTURES, COMPOUNDS AND SOLID SOLUTIONS

Up to now we have been considering the structure of a pure metal. But in industry we rarely have to deal with perfectly pure metals. Other elements are usually present, either as impurities which it is difficult to get rid of, or as deliberate additions to the basic metal, for the purpose of improving its properties. Any metal which is not a pure metallic element is called an *alloy*.

The introducing of another element into what was previously a pure metal may result in a mixture, a chemical compound, or a solid solution.

A *mixture* of two or more substances is such that each substance retains its own properties, and no chemical reaction occurs between them. An example of two metals which form a mixture is aluminium and lead. In many cases, whether two metals form a mixture or chemically combine is temperature-dependent; i.e., metals which do not combine at room temperature may combine at higher temperatures.

Often, metals combine together to form a compound which has different properties from either of the constituent metals. Such compounds are termed *intermetallic compounds*, examples being the compounds formed by aluminium and copper, or magnesium and silicon.

Many metals are completely soluble in the liquid state, and the liquid solution has entirely different properties from the two separate metals. In some cases the two metals are also completely soluble in the solid state in all proportions, and the metals have a common crystal lattice. They have then formed a *solid solution*. More commonly, metals are soluble in each other over a limited range of composition. Under the optical microscope, the structure of a solid solution will appear similar to that of a pure metal, since only one phase exists, with no evidence of the existence of two separate metals. On an atomic scale there are two main types of solid solution: (i) substitutional solid solutions and (ii) interstitial solid solutions.

Substitutional solid solutions may be of either the *ordered* or the *disordered* type. In the case of an ordered substitutional solid solution, some of the atoms of the parent metal (the *solvent*) are replaced in the crystal lattice by atoms of the second metal (the *solute*) in a regular pattern (Fig. 2.1(a)). In a disordered substitutional solid solution, atoms of the parent metal are replaced in the crystal lattice by atoms of the second metal in a random manner (Fig. 2.1(b)).

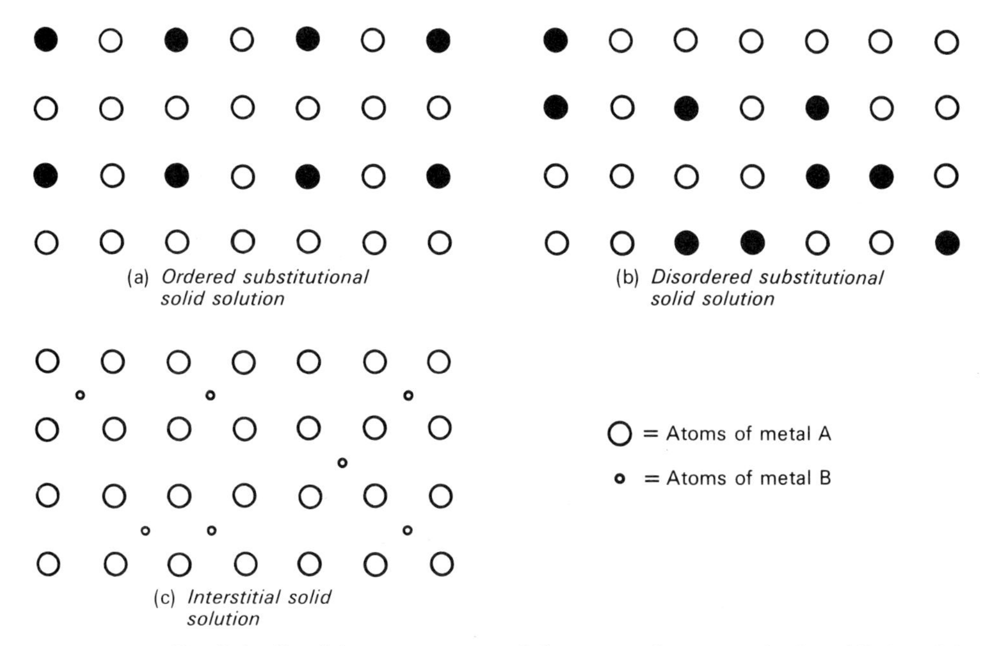

Fig. 2.1 Possible arrangements of the atoms of two metals, A and B, in solid solution

In some cases, solid solutions have the random structure at high temperatures and the ordered structure at low temperatures, and this change may result in a corresponding change in properties. For substitutional solid solutions to form, the atomic diameters of the two metals must be similar. An example of a substitutional solid solution is that formed by copper and nickel.

If the atomic diameter of one element is much smaller than that of the other, then the smaller atoms may fit into spaces in the crystal lattice of the parent element, and in this case an *interstitial solid solution* is formed. The best-known example of this type is carbon in iron (Fig. 2.1(c)).

THE DIFFERENT TYPES OF EQUILIBRIUM DIAGRAM

A *phase* may be defined as a part of a metal or alloy that is homogeneous – i.e., that has the same crystal structure and composition throughout. Thus a pure metal contains only a single phase, while an alloy of two metals may contain more than one phase.

We may have vapour, liquid or solid types of phases, but in this subject we are primarily concerned with solid phases.

In most cases of metallurgical interest, liquid metals are completely soluble in one another; i.e., they form a single liquid phase. When a mixture of liquid metals solidifies, however, there are four possibilities:

(i) The metals may be completely soluble in the solid state, and so form a single-phase structure.
(ii) The metals may be completely *in*soluble in the solid state.
(iii) The metals may be partially soluble in the solid state.
(iv) The metals may form compounds known as *intermetallic* or *intermediate compounds*.

The mechanical, thermal and electrical properties of an alloy will very much depend upon the nature of the solid phases formed during or after solidification.

We have seen in Chapter 1 that the cooling curve for the solidification of a pure metal results in a single thermal arrest – i.e., the metal solidifies at a constant temperature. While there are cases of alloys also solidifying at constant temperature, in most cases the cooling curve indicates that the alloy solidifies over a range of temperature. A typical experimental cooling curve for an alloy is shown in Fig. 2.2(a), and it will be noted that a specific temperature for the commencement of solidification and for the completion of solidification is not easily determined. For this reason the thermal arrests are often determined by plotting the temperature of the specimen against the time for the temperature to fall by a fixed small amount. Such a curve, which is known as an *inverse rate curve*, enables us to determine more accurately the temperature of the thermal arrests (Fig. 2.2(b)).

Thermal analysis (i.e. the plotting of cooling curves) is an important experimental technique which enables the metallurgist to predict the phase structures of a solid alloy, its properties, and therefore possible commercial applications. The rate of cooling used for thermal analysis studies is very slow; i.e., the alloy is cooled under 'equilibrium' conditions so that at any temperature during cooling, any change in composition is completed and the structure is stable. When an alloy system is studied, cooling curves are plotted for the two pure metals and for alloys of varying composition. The resulting data are presented in the form of an equilibrium or phase diagram with the composition scaled on the horizontal axis and temperature on the vertical axis. The composition may be presented as atomic percentage or (more usually) as percentage by weight of the constituent metals. While, strictly speaking, the resulting equilibrium diagram will only enable us to

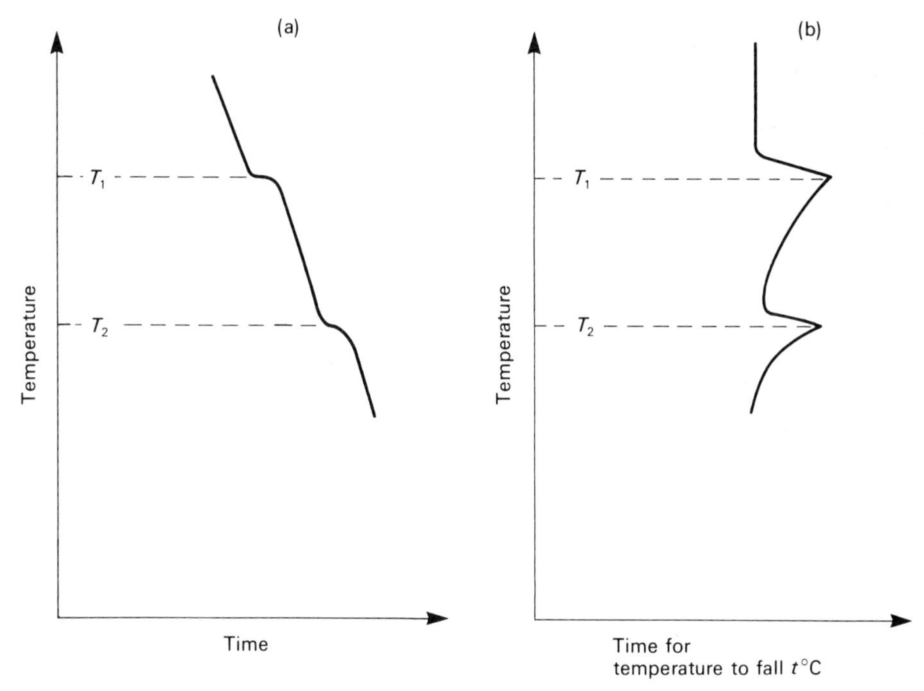

Fig. 2.2 (a) Normal cooling curve; (b) inverse rate curve

predict the structures and properties of alloys cooled under equilibrium conditions, it is possible, with experience, to predict from such a diagram the structures and properties to be obtained from the non-equilibrium cooling rates which occur in normal industrial practice.

THE SOLID-SOLUTION EQUILIBRIUM DIAGRAM

In some cases two metals are completely soluble in both the liquid state and solid state for all proportions of the constituent metals. Thermal analysis in this case will yield cooling curves for the two pure metals A and B, and for their alloys, as shown in Fig. 2.3. The *equilibrium diagram* (or *thermal equilibrium diagram*) is constructed by joining all the first thermal arrests (the temperatures at which solidification commences), and the resulting curve (the phase boundary) is termed the *liquidus*. By joining all the second thermal arrests, the temperatures at which solidification is complete, the *solidus* phase boundary is obtained. The resulting diagram is shown in Fig. 2.4, from which it will be seen that: above the liquidus only the liquid phase is present; below the solidus only solid phase is present; while between these two phase boundaries we always have a proportion of the liquid phase in equilibrium with a proportion of the solid phase. The region between the

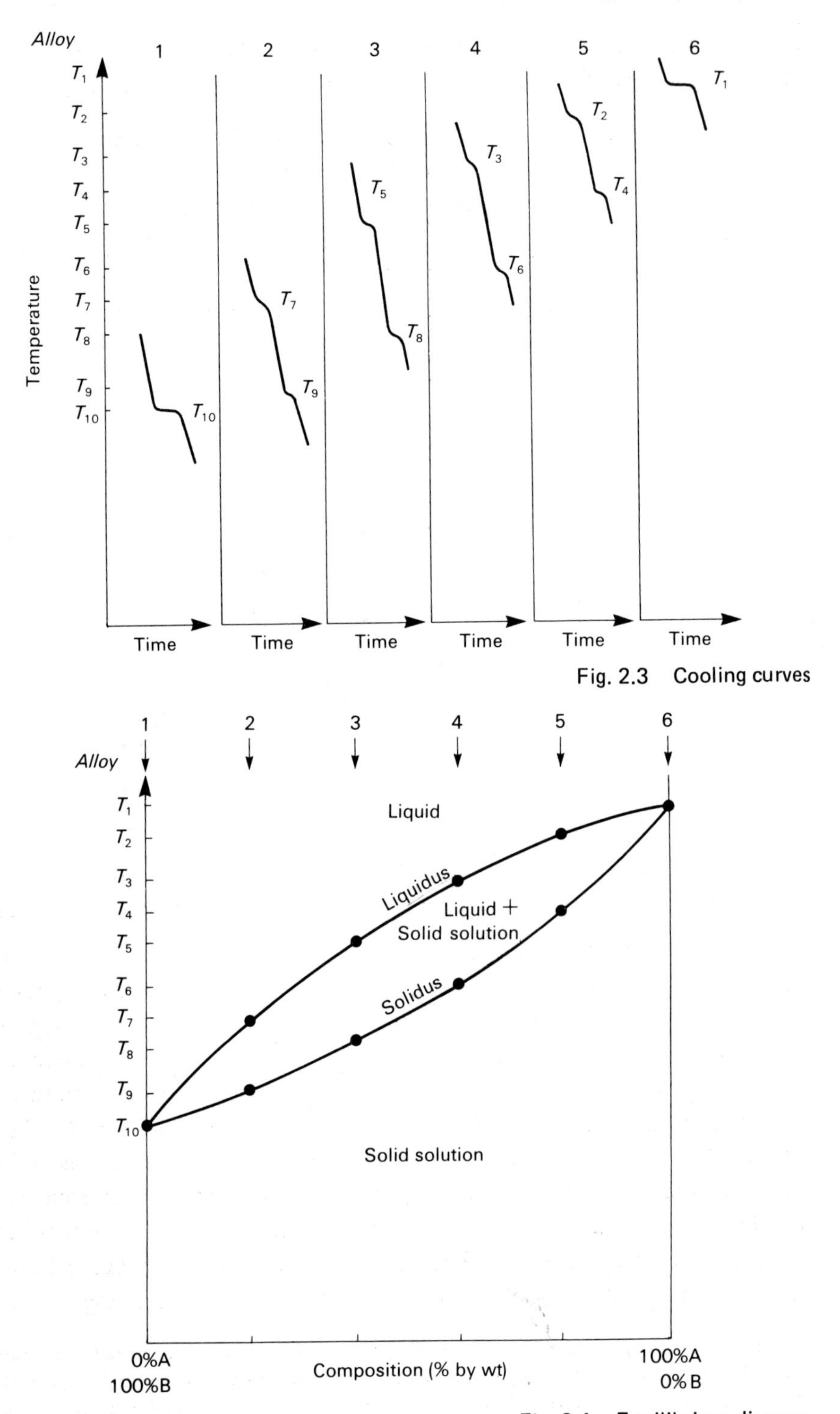

Fig. 2.3 Cooling curves

Fig. 2.4 Equilibrium diagram

liquidus and solidus is sometimes referred to as the *pastry range*. A diagram of this type is called a *solid-solution type of equilibrium diagram*.*

Fig. 2.5 shows how the physical properties of the alloy change as the proportion of one of the metals is varied from 0% to 100%. These curves apply to any alloy system in which the two metals are completely soluble in all proportions. We see that maximum hardness and tensile strength, and minimum electrical conductivity are obtained at about the 50% A–50% B alloy. This knowledge is important in selecting a particular alloy for a specific commercial application.

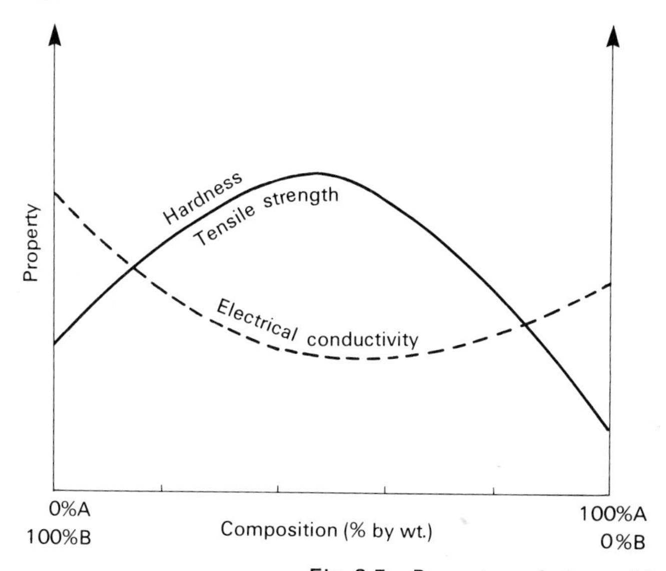

Fig. 2.5 Property variations with composition

We will now consider the information that may be derived from the equilibrium diagram concerning the composition and ratio of the phases present at any stage during the solidification of an alloy (Fig. 2.6). Taking, as an example, an alloy having the composition 60% A–40% B, we can deduce that the cooling curve for the alloy will be as shown in Fig. 2.7 – i.e., solidification commences at temperature T_1 and is complete at temperature T_4. Fig. 2.6 shows that at temperatures above T_1 the alloy would exist as a liquid solution, and at T_1, solid-solution crystals of a composition represented by *c* on the solidus would commence to solidify out of solution. We would thus have a liquid solution in equilibrium with a solid solution *c*. Since, however, these initial crystals (*dendrites*) are richer in A the remaining liquid will

*Note that the direction in which the curves slope depends on which metal we call A and which B. To avoid confusion, we shall make it a rule that the pure metal with the higher melting point shall be A, and that the scale of A shall go from 0% on the left to 100% on the right. Then all curves will slope the same way.

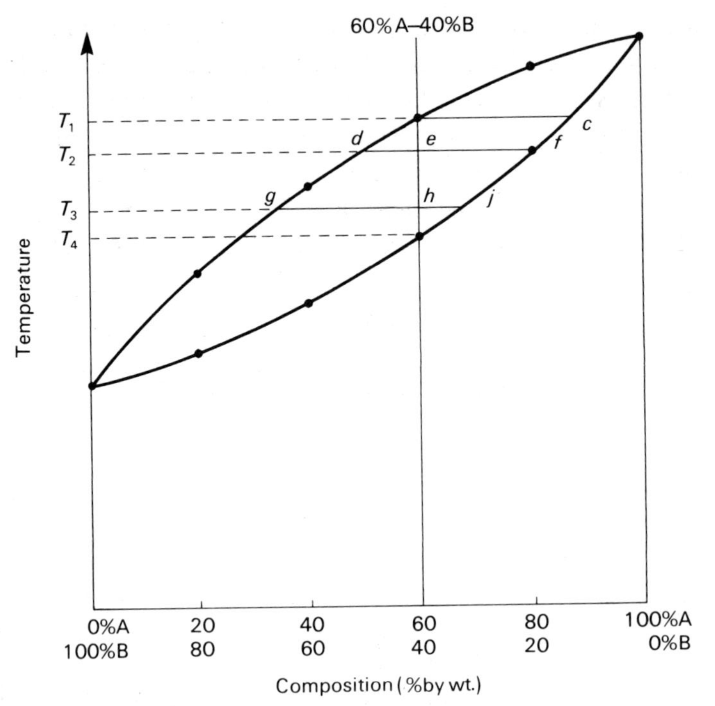

Fig. 2.6 Equilibrium diagram

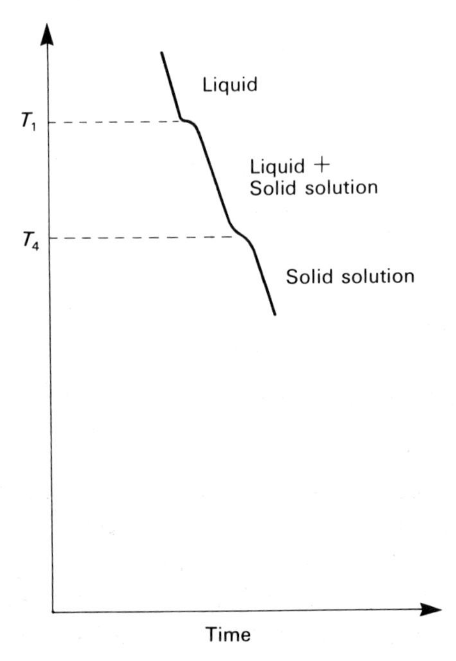

Fig. 2.7 Cooling curve for 60% A–40% B alloy

contain less A. Thus at a temperature of T_2 the liquid solution will have a composition represented by d and dendrites of composition f will solidify out of the melt. Since we are considering cooling under equilibrium conditions, solid solution that has previously solidified out between T_1 and T_2 will have assumed the composition represented by f by a process of diffusion. The structure of the melt will then be as shown in Fig. 2.8(a). Thus by drawing a horizontal line between the liquidus and solidus at any temperature we may specify the composition of liquid phase and solid phase present. That is, the point where the horizontal line meets the liquidus boundary indicates the composition of the liquid phase present, and the point where the line meets the solidus boundary gives the composition of the solid phase present.

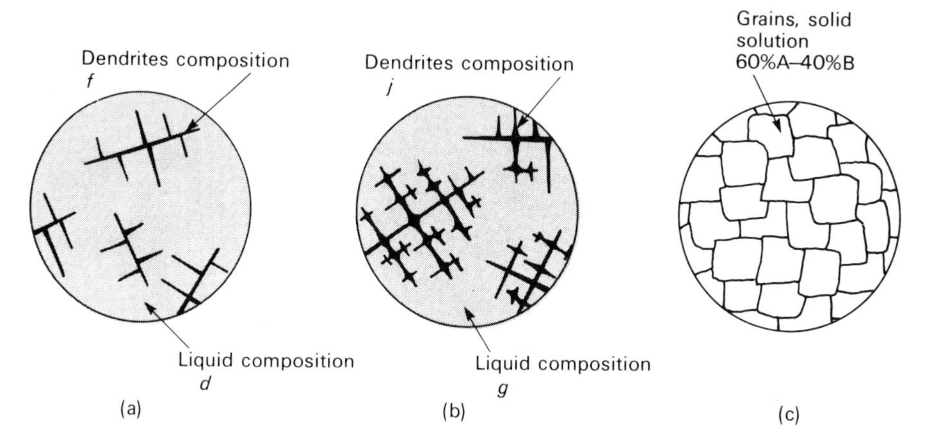

Fig. 2.8 Structures

Also, by considering the division of the horizontal line def into the parts de and ef it is possible to determine the ratio of solid to liquid phase present at T_2, i.e.

$$\text{Ratio}\left(\frac{\text{Solid phase } f}{\text{Liquid phase } d}\right) = \frac{\text{Length } de}{\text{Length } ef}$$

This is an inverse relationship known as the *lever rule*.

As solidification proceeds, the composition of the liquid present changes along the liquidus boundary and the composition of the solid present changes along the solidus boundary. Thus at temperature T_3, liquid solution of composition represented by g on the liquidus is in equilibrium with solid solution represented by j on the solidus, diffusion ensuring that these are the only two phases present and that they are in equilibrium. The structure of the melt at T_3 is shown in Fig. 2.8(b). By the lever rule,

$$\text{Ratio}\left(\frac{\text{Solid phase } j}{\text{Liquid phase } g}\right) = \frac{\text{Length } gh}{\text{Length } hj}$$

At temperature T_4 solidification is complete and the final solid structure will consist of a single-phase solid solution of composition 60% A–40% B. Note that the metals A and B are in solution, so nowhere does metal A or metal B exist on its own, any more than pure water or dry salt can exist in a salt solution. When examined under the microscope, the structure (Fig. 2.8(c)) will appear very similar to a pure metal structure, which is, of course, also a single-phase structure. The grains shown in Fig. 2.8(c) will be of a fairly uniform size and shape and are referred to as *equi-axed grains*.

The best-known example of two metals which exhibit a complete range of solid solubility is copper and nickel, and the copper–nickel equilibrium diagram is shown in Fig. 2.9. Copper–nickel alloys are of considerable commercial importance, having excellent corrosion resistance, and they are used in such applications as condenser tubes, coinage and electrical resistors. Commercial alloys such as Monel have additions of other elements to improve the properties of the alloys for specific applications. Typical additional alloying elements used are iron, silicon, manganese, and aluminium, usually as additions of the order of 0–3%. The addition of these alloying elements

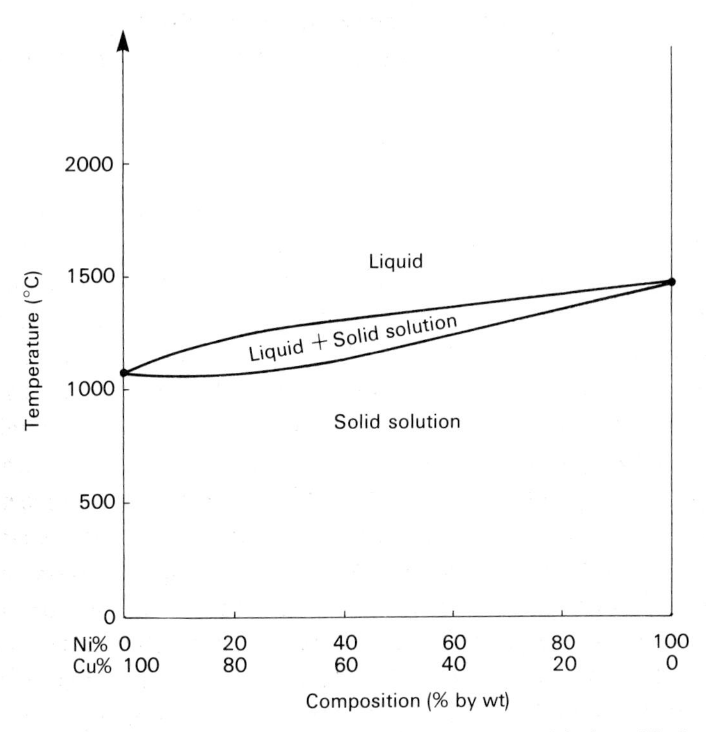

Fig. 2.9 Copper–nickel equilibrium diagram

can result in significant improvements in tensile strength, often obtained by heat treatment. Single-phase solid-solution alloys will not respond to hardening types of heat treatment. These are only effective when additional phases are introduced into the structure by these further alloying elements in commercial alloys.

SELF-TEST QUESTION 1

Cooling curves from alloys consisting of various proportions of two metals A and B gave the following results:

% A	0	10	20	30	50	60	80	90	100
1st arrest (°C)	600	740	860	960	1140	1220	1320	1370	1400
2nd arrest (°C)	–	630	690	760	910	1000	1160	1280	–

(a) Plot and label the equilibrium diagram.

(b) For an alloy containing 40% of A and 60% of B state:

- (i) the temperature at which solidification commences;
- (ii) the temperature at which solidification is completed;
- (iii) the compositions of the phases present at 900 °C;
- (iv) the ratio of the phases present at 900 °C. (*Solution, p. 165*)

CORED SOLID-SOLUTION STRUCTURES

We now have seen what happens when a liquid solution of two metals is cooled slowly, under equilibrium conditions, giving time for the dendrites to change in composition by diffusion. Under most practical conditions of rapid cooling, however, and certainly under commercial casting conditions, the cooling rate is much faster, and there is therefore insufficient time for composition adjustments by diffusion to occur. The result of this is that (referring to Fig. 2.6) at T_2 the dendrites present in the melting will have a layered (cored) structure, the centre of the dendrite having a composition rich in metal A, and subsequent layers having compositions between c and f. The average composition of the solid dendrites present will be richer in A than is represented by f on the diagram, and since the total composition of the (solid + liquid) present remains constant the liquid solution present at T_2 will be correspondingly deficient in A (and hence richer in B). In the same way, at T_3 the solid dendrites will have a composition richer in A than is represented by j and the liquid solution will be correspondingly more deficient in A than is represented by g on the diagram. The final structure when solidification is complete will therefore be as shown in Fig. 2.10(a) and for the particular case of a copper–nickel alloy will be as shown in Fig. 2.10(b). The variation in composition of the cored dendrites is observed under the microscope by colour variations across the dendrites. This cored structure is undesirable, since it results in a non-homogeneous structure which will result in unpredictable variations in physical properties. The structure may be converted to the equilibrium structure of the type shown in Fig. 2.8(c) by annealing the alloy at a temperature just below the solidus temperature, allowing diffusion to take place with resulting adjustments in composition. The process of diffusion is both temperature-dependent and time-dependent, and therefore it is necessary to use a correct combination of annealing temperature and time.

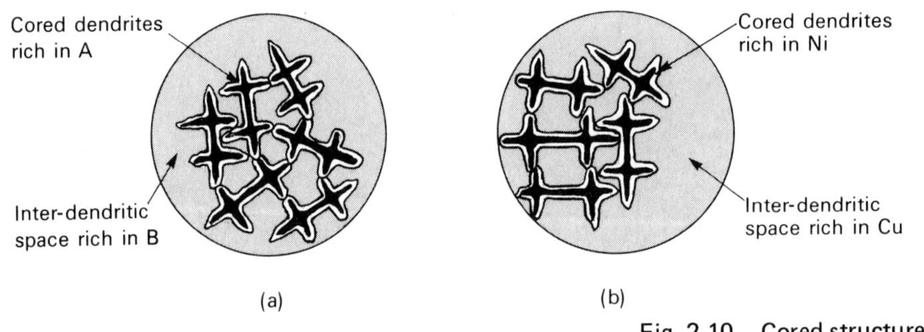

Fig. 2.10 Cored structures

TWO METALS COMPLETELY INSOLUBLE IN THE SOLID STATE (EUTECTIC TYPE)

This case occurs when two metals are completely soluble in all proportions in the liquid state but completely insoluble in the solid state. As before, the equilibrium diagram may be constructed from the results of thermal analysis experiments. Typical cooling curves and the resulting equilibrium diagram are shown in Figs. 2.11 and 2.12. The significant features of Fig. 2.12 are:

(i) that the two parts of the liquidus phase boundary meet at E (known as the *eutectic point*, the composition at this point being known as the *eutectic alloy*);

(ii) apart from the two pure metals, all alloys have a common second thermal arrest temperature T_E known as the *eutectic temperature*. The lines joining CE and ED are the liquidus phase boundaries and the solidus phase boundary is CFEGD.

Referring to Fig. 2.13 (which is a repetition of Fig. 2.12) and considering the

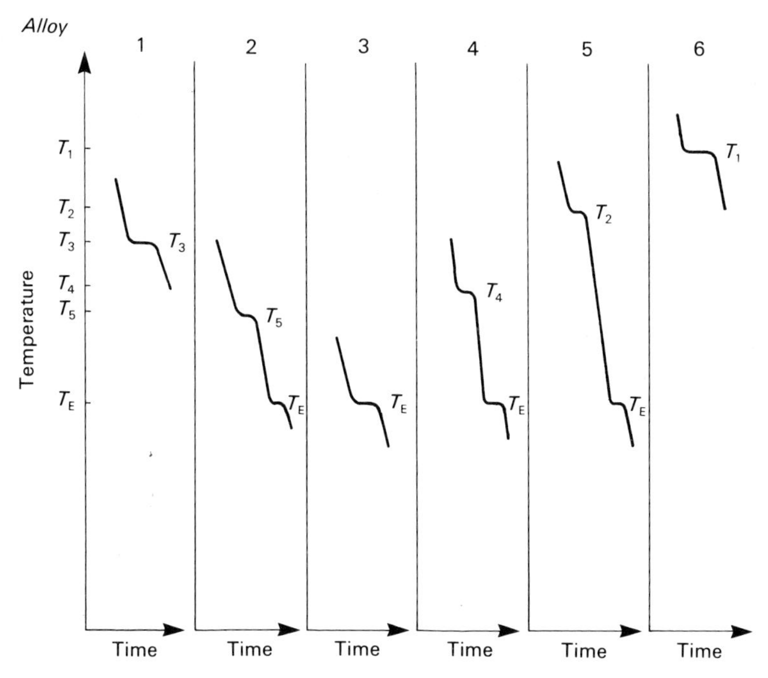

Fig. 2.11 Cooling curves

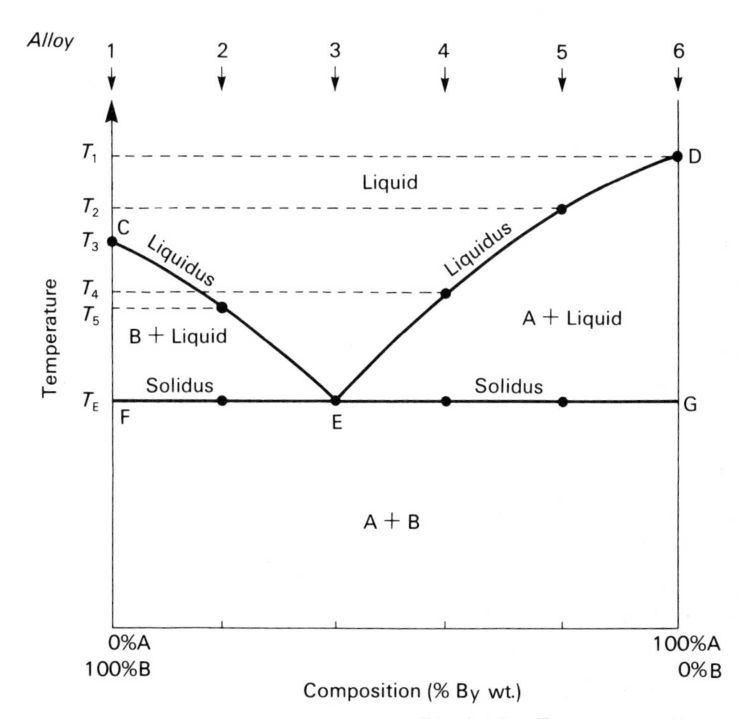

Fig. 2.12 Eutectic equilibrium diagram

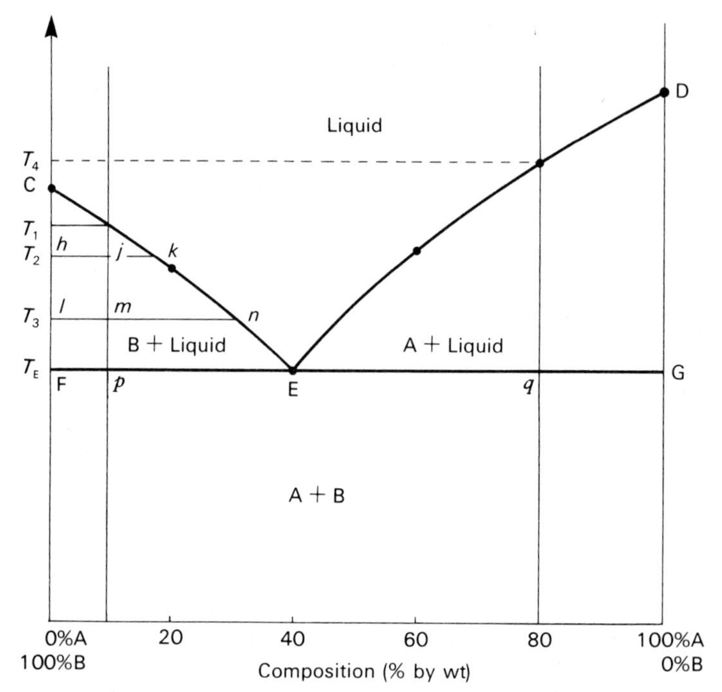

Fig. 2.13 Eutectic diagram

solidification of the 10% A–90% B alloy, we see that solidification commences at temperature T_1 when dendrites of pure metal B solidify out and consequently the liquid is deficient in B. (In general, the first phase to solidify out of a melt is called the *primary phase*.) As the temperature falls to some temperature such as T_2, dendrites of pure metal B continue to solidify out from the melt. By applying the equilibrium diagram rules we have already met, we can use the horizontal line hjk at T_2 to determine both the composition and ratio of phases present, the solid phase being represented by h on the diagram (pure metal B) and the liquid phase by k. Applying the lever rule gives

$$\text{Ratio}\left(\frac{\text{Solid pure metal B}}{\text{Liquid solution } k}\right) = \frac{jk}{hj}$$

As the temperature continues to fall, further dendrites of pure metal B solidify out and the liquid solution changes in composition down the liquidus boundary. Thus at temperature T_3 we have pure metal B in equilibrium with a liquid solution represented by n, the ratio of the two phases being

$$\text{Ratio}\left(\frac{\text{Solid pure metal B}}{\text{Liquid solution } n}\right) = \frac{mn}{lm}$$

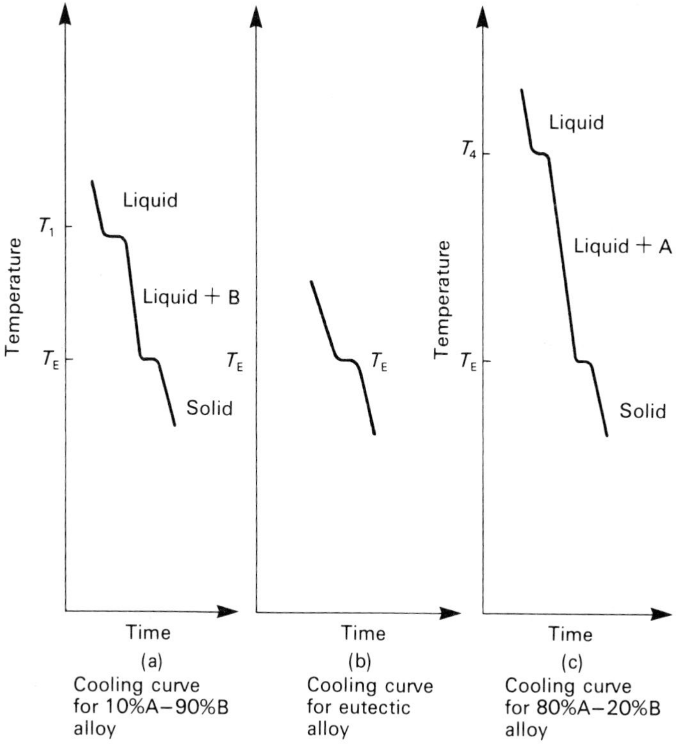

Fig. 2.14 Cooling curves

The temperature T_E represents the lowest point at which any liquid may exist. At this temperature the remaining liquid, of composition E, solidifies by alternately crystallising out of solution into pure metal B and pure metal A until solidification is complete. The final solid structure will therefore consist of grains of pure metal B (termed the *primary phase*) surrounded by a mixture of pure metal A and pure metal B (known as the *eutectic mixture*). Applying the lever rule for the case of complete solidification, we have

$$\text{Ratio}\left(\frac{\text{Primary pure metal B}}{\text{Eutectic mixture of B and A}}\right) = \frac{\text{E}p}{\text{F}p}$$

The ratio of pure metal B to pure metal A in the eutectic mixture may be determined by applying the lever rule for the line FEG; i.e. in the eutectic mixture

$$\text{Ratio}\left(\frac{\text{Pure metal B}}{\text{Pure metal A}}\right) = \frac{\text{EG}}{\text{EF}}$$

The structures during the process of solidification of the alloy and the final room temperature structure are shown in Fig. 2.15. The final structure consists of primary grains of pure metal B and the eutectic mixture of pure

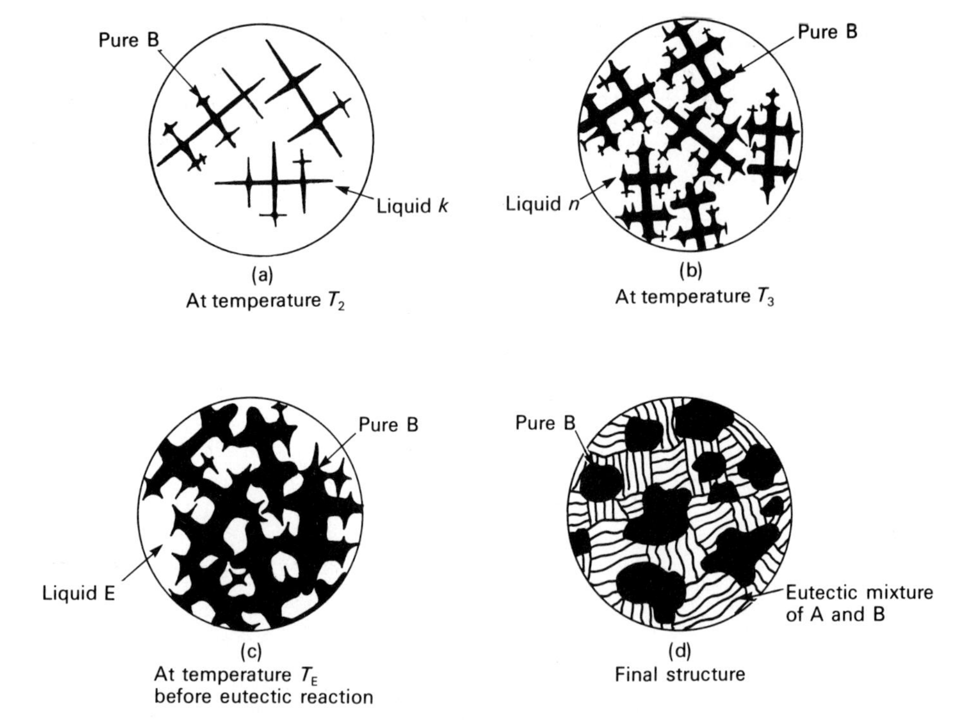

Fig. 2.15 Structures

A and pure B which may appear as a *lamellar* structure (i.e. a plate-like structure) due to the process of solidification of the eutectic liquid. Alternatively, the eutectic mixture may appear in the final structure as *spheroidal* particles (i.e. rounded particles).

Referring again to Fig. 2.13, we see that any alloy between E and G will solidify by ejecting pure metal A during cooling until T_E is attained when the eutectic liquid will again solidify as a mixture of the two pure metals A and B. In this case the final structure will consist of primary A and the eutectic mixture; e.g., for the 80% A–20% B alloy the final structure will consist of primary A and the eutectic mixture in the ratio

$$\frac{\text{Pure metal A}}{\text{Eutectic mixture}} = \frac{\mathrm{E}q}{\mathrm{G}q}$$

As before, the eutectic mixture will consist of A and B in the ratio

$$\frac{\text{Pure B}}{\text{Pure A}} = \frac{\mathrm{EG}}{\mathrm{EF}}$$

Considering the particular case of an alloy of composition E (i.e. the eutectic alloy) it will be seen by reference to Fig. 2.11 and 2.14(b) that this alloy will exhibit a single thermal arrest at T_E; i.e., it solidifies at a constant temperature as if it were pure metal, and therefore shows no pasty range. In this particular case the final structure will contain only the eutectic mixture, as in Figs. 2.16 and 2.17. The variation in hardness, tensile strength and electrical conductivity with composition is shown in Fig. 2.18

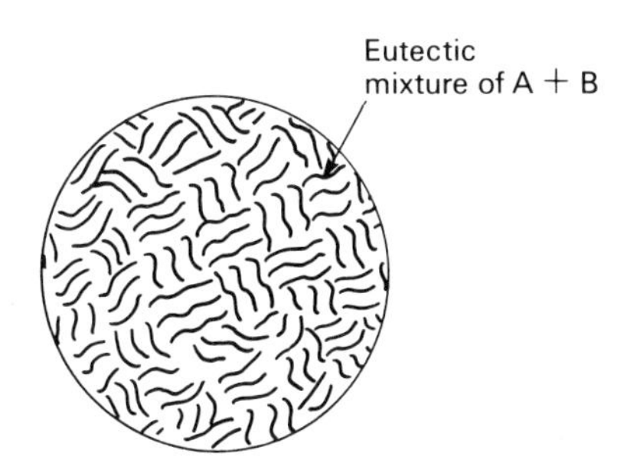

Fig. 2.16 Eutectic alloy structure

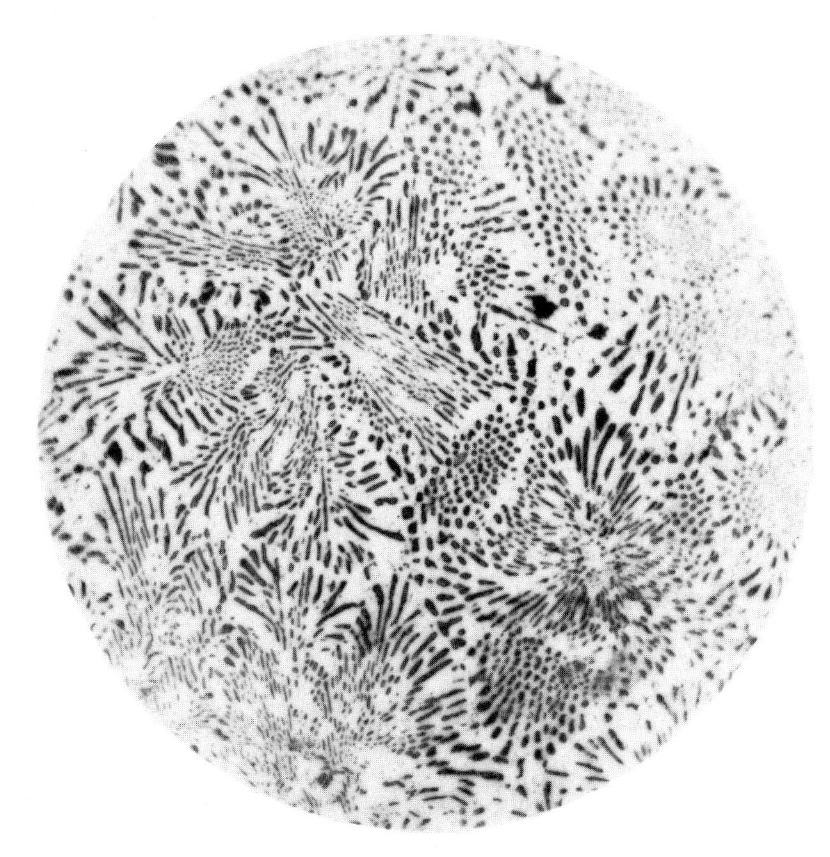

Fig. 2.17 Example of a eutectic structure – silver solder eutectic (magnification x 400)

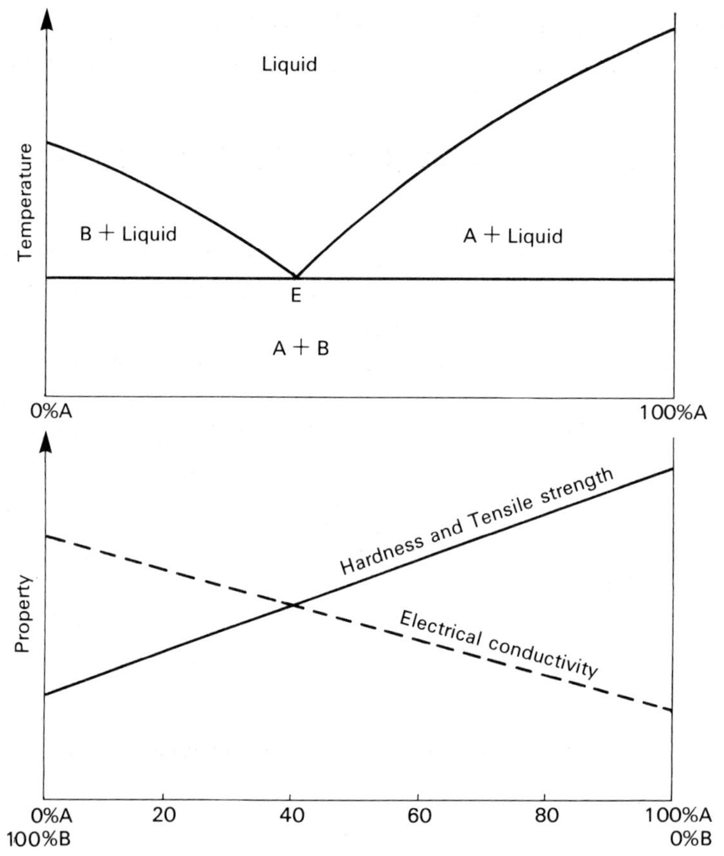

Fig. 2.18 Variation of properties with composition

SELF-TEST QUESTION 2

From the cooling curves of various alloys of zinc and cadmium the following results were obtained:

% Cadmium	0	20	40	60	86	90	100
1st arrest (°C)	419	382	345	308	266	290	321
2nd arrest (°C)	—	266	266	266	—	266	—

Using the above data, draw and label the equilibrium diagram for this series of alloys.

With reference to a cooling curve, describe the cooling of an alloy containing 30% cadmium, estimating:

(a) the composition of the constituents present at 320°C;

(b) the ratio, by weight, of solid to liquid at 320 °C;

(c) the proportion of eutectic in the final structure. (*Solution, p. 166*)

THE PARTIAL-SOLUBILITY (COMBINATION) EQUILIBRIUM DIAGRAM

In many alloy systems two metals are soluble in the solid state over a limited range of composition and a eutectic is formed between two solutions. This is illustrated by the lead–tin diagram shown in Fig. 2.19. The symbol α (alpha) is used to represent lead-rich solid solutions and β (beta) represents tin-rich solid solutions. (In general the dividing line, on an equilibrium diagram, between two different *solid* phases is called a *solvus*.) The composition of the eutectic alloy is 62% tin–38% lead, and the eutectic temperature 183 °C. At the eutectic temperature a eutectic mixture of a 20% solution of tin in lead and a 3% solution of lead in tin is formed, and resulting structures will consist of the primary solid solutions and a eutectic mixture of these two solid solutions. The diagram also illustrates another feature to be covered in the next section – i.e., the phase boundary *bc* indicates that the solubility

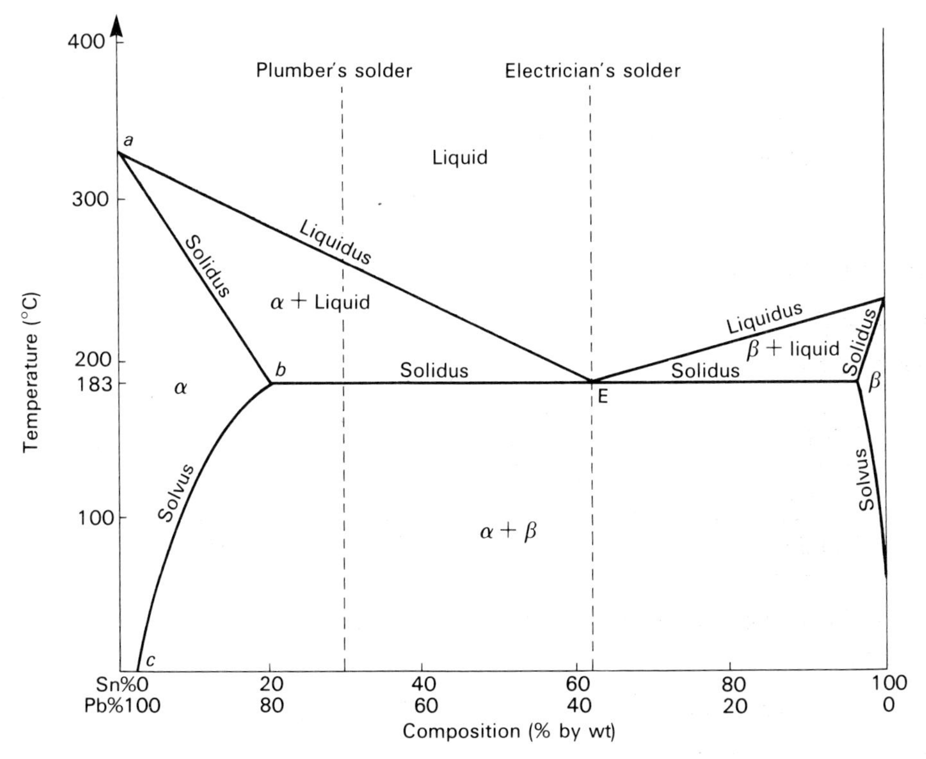

Fig. 2.19 Equilibrium diagram for lead and tin – an example of a partial-solubility equilibrium diagram

of tin in lead changes in the solid state, since at 183 °C, 20% tin is soluble in lead, while at room temperature only 3% tin is soluble. The lead–tin alloys are the basis of many commercial soft solders, the two most significant being *plumber's solder*, and *electrician's solder*. The 70% lead solder known as 'plumber's solder' shows a wide pasty range allowing ample time for 'wiping' the joint. The eutectic alloy solder consisting of 62% tin–38% lead is often known as *tinman's solder* or *electrician's solder* since this alloy will show no pasty range and is therefore useful for making electrical connections.

Fig. 2.20 shows how the hardness, tensile strength and electrical conductivity of lead–tin alloys vary when the composition of the alloy is varied.

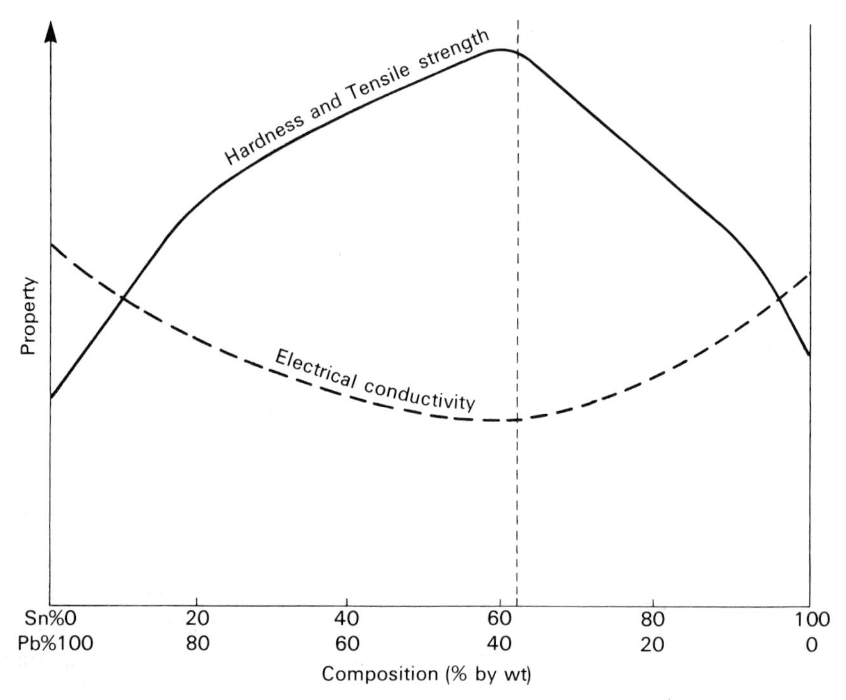

Fig. 2.20 Variation of hardness, tensile strength, and electrical conductivity with percentage composition of lead–tin alloys

CHANGE OF SOLUBILITY IN THE SOLID STATE, AND PRECIPITATION HARDENING

The B-rich portion of the equilibrium diagram for two metals A and B is shown in Fig. 2.21, and it will be seen that whereas at 500 °C, 15% A is soluble in B, at 0 °C only 5% A is soluble in B. Considering, then, the 10% A alloy, at 340 °C all the metal A will be in solution, but on further slow cooling to 0 °C, since only 5% A may be retained in solution, the excess 5% A

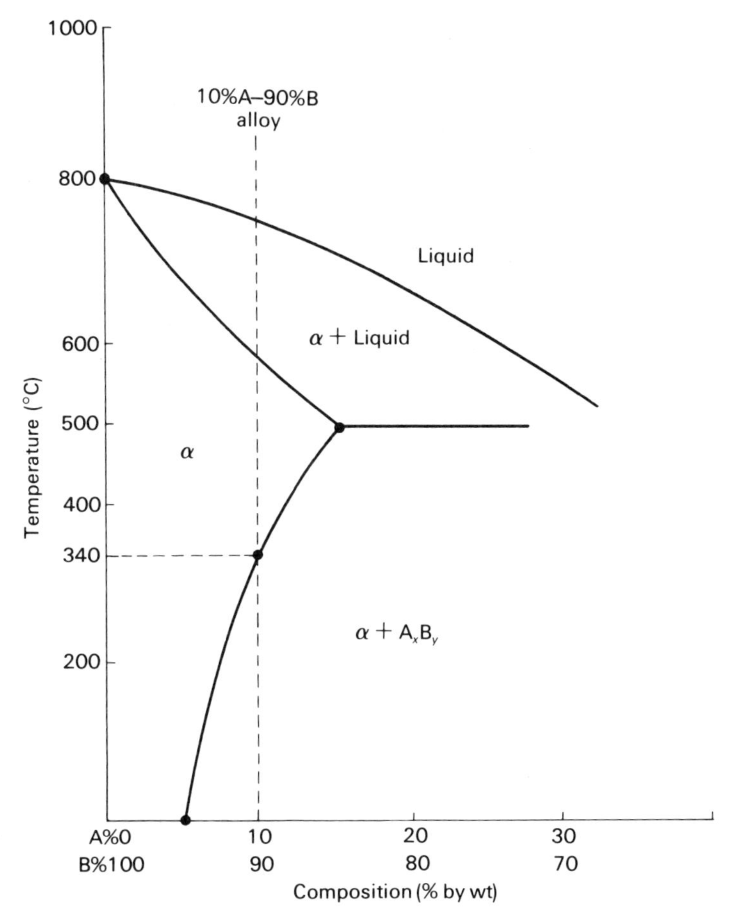

Fig. 2.21 Partial-solubility equilibrium diagram to which the grain structures of Fig. 2.22 relate

is precipitated out of solution in the form of an intermetallic compound A_xB_y. The precipitated compound may be present in the final structure either within the grains or in the grain boundary region of the solid solution. The slow cooling of the alloy from above 340°C to give the structure shown in Fig. 2.22(b) represents an annealing operation, and in this condition the alloy would be soft and ductile.

If, however, the alloy is rapidly cooled from above 340°C, the precipitation of the excess A in the form of the intermetallic compound is suppressed, and the 10% A is retained in solid solution. This is known as a *supersaturated solid solution*. In this condition the alloy will have properties rather similar to the annealed condition. In order to obtain this structure it is often necessary to water-quench the alloy and this treatment is called *solution treatment*. The structure is unstable, and there will be a strong tendency for

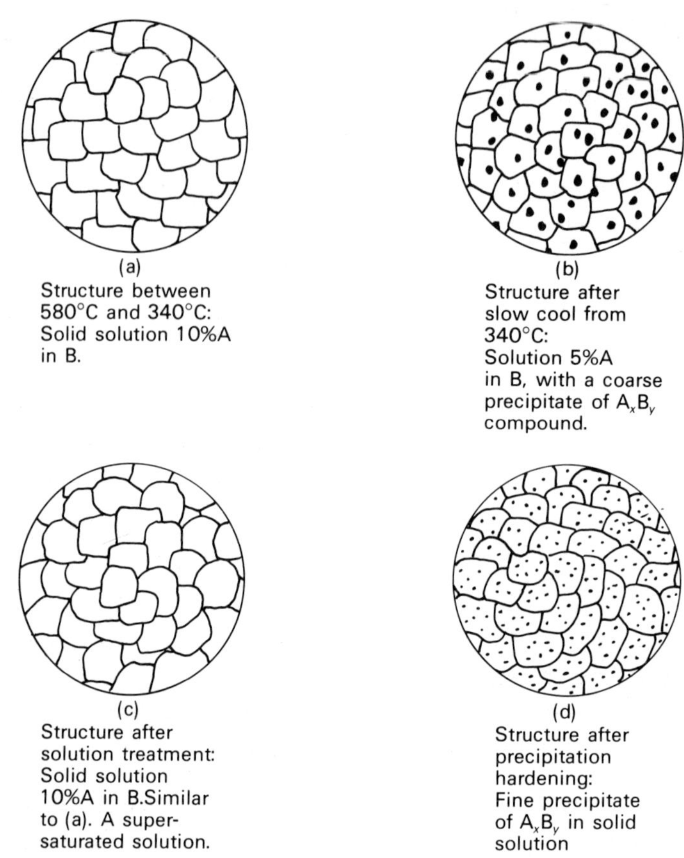

Fig. 2.22 Grain structures of the 10% A–90% B alloy of Fig. 2.21 after various treatments

the excess A in solution to precipitate out naturally at room temperature, to attain the equilibrium state of only 5% A being retained in solid solution. In many cases it is necessary to trigger the precipitation mechanism by a low-temperature treatment known as *precipitation hardening*. After either the natural precipitation or the precipitation heat treatment, the excess A will be precipitated in the form of the compound $A_x B_y$ in a very fine form often not visible under the optical microscope. Associated with this very fine precipitate is a marked increase in hardness and tensile strength.

Alloys that exhibit the solution-treatment and precipitation-hardening mechanism are very important commercially, probably the best-known example being the Duralumin-type alloys, based on the Al–4% Cu alloy used in the aircraft industry. The advantage of this type of alloy is that it may be readily formed in the solution-treated condition, and after forming, only a low-temperature precipitation-hardening treatment is required, to obtain high mechanical properties.

SUMMARY

- A *mixture* of two or more substances may be in any proportions and each substance retains its own properties.
- The constituents of a *compound* combine in one fixed proportion and the properties of the constituents are replaced by the properties of the compound.
- A substance consisting of two or more elements, of which at least one is a metal, is termed an *alloy*.
- When a metal is in solid solution in another metal, its atoms fit into the space lattice of the other metal, without unduly distorting it. If they replace atoms of the other metal in the space lattice, they form a *substitutional solid solution*, which may be either *ordered* or *disordered*. If they are small enough to fit in between the atoms of the other metal, they form an *interstitial solid solution.*
- In a *phase* of a metal or alloy, the material has the same crystal structure and composition throughout, if it is a solid. Alternatively, the material may be in the *liquid phase* or the *vapour phase*. A *primary phase* is the first solid phase to appear when a molten alloy is cooled.
- When an alloy is in *equilibrium*, any change in its structure due to heating or cooling has been completed, and the material is *stable*.
- An *equilibrium diagram* is a diagram with a vertical scale of temperature and a horizontal scale of percentage composition of constituents. It shows the limits of temperature and composition within which the various phases of an alloy are stable.
- A *liquidus* is a line on an equilibrium diagram, defining the temperatures at which solidification commences.
- A *solidus* is the corresponding line for temperatures at which solidification is just complete.
- A *solvus* is the corresponding line for temperatures at which one solid phase changes into a different solid phase.
- The *eutectic temperature* is the lowest temperature at which a mixture of two or more insoluble substances can remain liquid. It is lower than the solidification temperatures of any of the pure constituents. The *eutectic point* on an equilibrium diagram is the point whose coordinates are the composition of the mixture which forms the eutectic, and the

eutectic temperature. At the eutectic temperature, when an insoluble mixture is being cooled, simultaneous crystallisation of the remaining liquid into its constituents occurs.

- An equilibrium diagram is constructed from the thermal arrest points on cooling curves. Each cooling curve is a graph of temperature against time, plotted when a particular solution of metals cools from the liquid state to room temperature. Enough cooling curves are plotted to establish the equilibrium diagram over the full range of proportions, from 0% to 100%, of the constituents of an alloy.

- Molten metals can be taken as completely soluble in each other in all proportions, unless otherwise stated. When a solution of liquid metals solidifies, the metals may be completely soluble in each other and so form a solid solution, or they may be partly soluble, or they may be completely insoluble, or they may combine to form intermetallic compounds.

- When the metals are soluble in the solid state in all proportions there is no eutectic, and the equilibrium diagram has the form shown in Fig. 2.23(a). When the metals are completely insoluble, it has the form shown in Fig. 2.23(b). When the metals are soluble in each other over a limited range of proportions it has the form shown in Fig. 2.23(c), in which the shaded areas represent solid solutions of metal A in metal B, and metal B in metal A, respectively, while between them, at temperatures below the eutectic temperature, is a eutectic mixture of the two solid solutions.

- The equilibrium diagram provides a forecast of the phase or phases which will be present when an alloy of a given composition is heated or cooled to a specific temperature under equilibrium conditions (i.e. sufficiently slowly). Except in the case of a solid solution, there will be a mixture of the two phases which occur on either side of the point representing the given temperature and composition. These two phases will be in proportions which are inversely proportional to the horizontal distances to the corresponding phase boundaries (the *lever rule*).

- It follows, from the lever rule, that when dendrites first start to solidify from a melt of two metals which form a solid solution, the dendrites are richer in one metal and poorer in the other metal than the overall composition of the mixture, while the remaining liquid is correspondingly richer and poorer in the opposite direction. If the cooling is slow enough, the proportions of the metals in the dendrites and in the liquid adjust by diffusion, so that when solidification is complete, we have a homogeneous solid solution in the original proportions. If the cooling is too quick, however, (e.g. as in normal casting) there is not enough time for this, so the final grain structure consists of dendrites richest in one metal at their centres, and in between, solid solution rich in the other metal. This effect is called *coring*. It is eliminated by annealing at just below solidus temperature, until diffusion has made the composition of the material uniform throughout.

- *Precipitation hardening* applies to cooling through the solvus from one of the shaded areas of Fig. 2.23(c), as indicated by the broken line. If the cooling is slow, the reduced solubility at room temperature produces a coarse precipitate of intermetallic compound, giving a soft ductile material. If the cooling is rapid, however, (e.g. quenching) a soft supersaturated solid solution results. In time, a very fine precipitate forms, giving increased hardness. This effect can be brought about quickly by a low-temperature heat treatment.

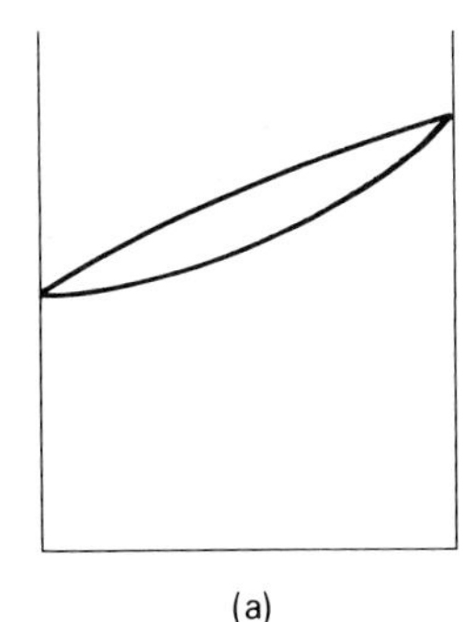

(a)

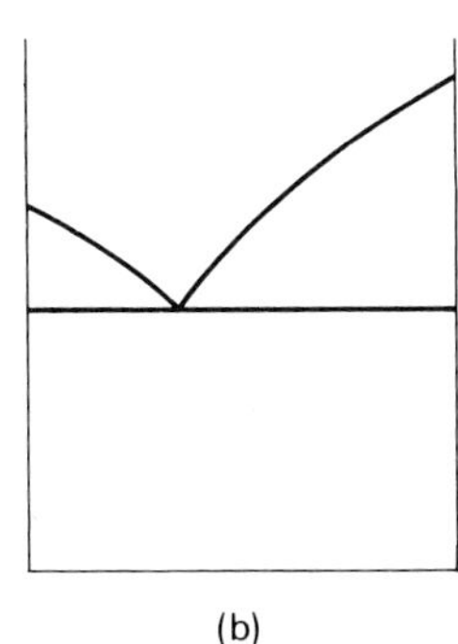

(b)

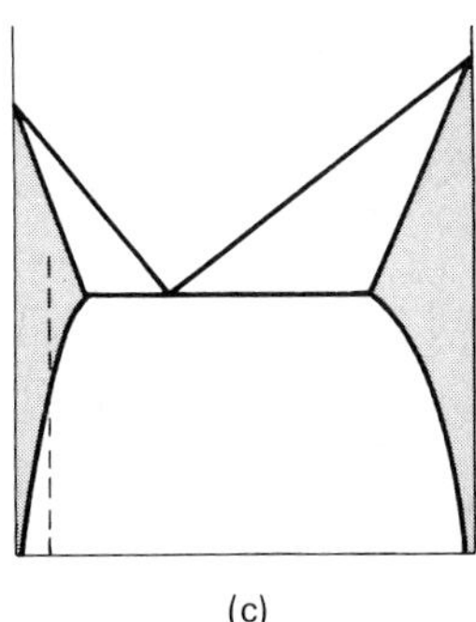

(c)

Fig. 2.23 Three equilibrium diagrams

EXERCISE 2

1) Explain the differences between a mixture, a compound and a solid solution.

2) Write definitions of the following terms:

(a) an alloy;

(b) latent heat of fusion;

(c) thermal arrest;

(d) equilibrium (as applied to a solidifying alloy);

(e) a phase.

3) Explain, with the aid of diagrams, the differences between:

(a) an ordered substitutional solid solution;

(b) a disordered substitutional solid solution;

(c) an interstitial solid solution.

4) Sketch: (a) the cooling curve for an alloy showing two thermal arrests; and (b) the corresponding inverse rate curve. (c) State the method of obtaining an inverse rate curve, and explain why this type of curve may be preferable.

5) (a) Describe in detail, how the equilibrium diagram for two given metals A and B would be determined experimentally.

(b) Sketch the typical form of such a diagram, if the two metals are: (i) completely soluble, (ii) completely insoluble in the solid state. Indicate on your sketches the liquidus, the solidus, and the eutectic point (where applicable).

6) Sketch graphs showing the variation in tensile strength and electrical conductivity as the percentage composition is varied, for alloys of two metals which are: (a) completely soluble, (b) completely insoluble, (c) partially soluble, in the solid state.

7) Distinguish clearly between the following terms, using diagrams where necessary:

(a) face-centred cubic lattice and body-centred cubic lattice;

(b) substitutional and interstitial solid solution;

(c) liquidus, solidus and solvus.

8) (a) Sketch the three main metallic unit cells.

(b) Determine the number of atoms associated with each unit cell.

(c) State the relationship between the type of unit cell and the ductility of the metal.

(d) Explain the meaning of: (i) an ordered substitutional solid solution; (ii) an interstitial solid solution.

9) From cooling curves of various alloys of bismuth and antimony the following results were obtained:

% Antimony	0	20	40	60	80	100
1st arrest (°C)	271	400	490	550	600	631
2nd arrest (°C)	–	285	320	370	450	–

(a) Using the above data, plot and label the bismuth–antimony thermal equilibrium diagram.

(b) Describe, with reference to the cooling curve, the cooling of an alloy containing 50% of each metal, estimating:

(i) the temperature at which solidification commences;
(ii) the temperature at which solidification is completed;
(iii) the compositions of the liquid and solid phases at 420 °C;
(iv) the ratio of the phases present at 420 °C.

(c) What would be the effect of rapid cooling on the microstructure of the alloy?

10) Thermal analysis of two metals A and B gave the following results:

% A	0	10	20	30	40	50	60	70	80	90	100
1st arrest (°C)	700	870	1000	1120	1220	1300	1365	1420	1450	1480	1500
2nd arrest (°C)	–	730	800	860	950	1040	1140	1240	1340	1430	–

(a) Draw the equilibrium diagram and label the phase fields.

(b) For the 25% A alloy state: (i) the composition of the phases in equilibrium at 1000 °C; (ii) the ratio of the phases present at 1000 °C, assuming an equilibrium rate of cooling.

(c) Sketch and label the room-temperature structure of the 25% A alloy: (i) cooled under equilibrium conditions; (ii) cooled at a faster rate than the equilibrium rate.

(d) For the whole range of alloys from 0% A to 100% A sketch a diagram to show how the tensile strength is likely to vary with changes in composition.

11) From the cooling curves of various alloys of two metals A and B, the following results were obtained:

Percentage of A	0	10	20	45	60	80	100
Percentage of B	100	90	80	55	40	20	0
1st arrest (°C)	1100	980	870	800	960	1180	1400
2nd arrest (°C)	–	700	700	700	700	700	–

Using the above data, draw and label the thermal equilibrium diagram for this series of alloys.

With reference to a cooling curve, describe the cooling of an alloy consisting of 70% A and 30% B, estimating:

(a) the composition of the constituents present at 900 °C;

(b) the ratio, by weight, of solid to liquid at 900 °C;

(c) the percentage of eutectic in the final grain structure;

(d) the composition of the eutectic.

12) From the cooling curves of various alloys of cadmium and bismuth, the following results were obtained:

% Cadmium	0	10	25	40	55	70	85	100
1st arrest (°C)	268	238	191	140	213	263	297	321
2nd arrest (°C)	–	140	140	–	140	140	140	–

Using the above data, draw and label the equilibrium diagram for this series of alloys.

By means of the equilibrium diagram, determine for an alloy consisting of 20% cadmium, 80% bismuth:

(a) the temperature at which it starts to solidify;

(b) the composition of the constituents present at 180 °C;

(c) the percentages of solid and liquid present at 180 °C by weight;

(d) the percentage of eutectic present in the grain structure when it has cooled.

13) Bismuth melts at 271 °C and tin at 232 °C. They form a eutectic containing 44% tin which melts at 132 °C. Bismuth dissolves a maximum of 4% tin, and tin a maximum of 12% bismuth at the eutectic temperature. Draw and label the bismuth–tin thermal equilibrium diagram. Describe, with reference to cooling curves, the cooling of alloys containing (a) 60% tin, (b) 90% tin from the liquid state to room temperature.

14) (a) Sketch and label the lead–tin equilibrium diagram.

(b) Explain why electrician's or tinman's solder has the composition of the eutectic alloy, whereas plumber's solder has a composition with more lead and less tin.

15) Thermal analysis of two metals A and B, gave the following results:

% A	0	5	20	55	70	90	100
% B	100	95	80	45	30	10	0
1st arrest (°C)	1300	1210	950	850	1100	1430	1600
2nd arrest (°C)	–	950	600	600	600	1100	–

(a) Draw and label the equilibrium diagram.

(b) State the composition of the eutectic alloy.

(c) State the maximum solubility of A in B.

(d) State the maximum solubility of B in A.

16) (a) With special reference to the Al–4% Cu alloy system, explain the meaning of the terms: (i) solution treatment; (ii) precipitation hardening.

(b) Outline a typical process schedule for a component produced from an Al–4% Cu type alloy, assuming that the manufacturing process includes a severe forming operation and that maximum tensile strength is required in the finished component.

17) What is meant by the term *coring*, as applied to the grain structure of an alloy? How can it be recognised, what is its cause, how can it be eliminated from an alloy in which it has occurred, and what takes place during the treatment for its elimination?

Why is coring undesirable?

THE EFFECTS OF CARBON IN STEEL

INTRODUCTION

In this chapter and in Chapters 4 and 5, we shall be considering the very important range of alloys of iron and carbon which go under the general names of:

steel, for carbon contents varying from 0% to about 1.3% by weight; and

cast iron, for carbon contents varying between about 2% and 4% by weight.

The equilibrium diagram for the full range of carbon contents from 0% to 7% C is shown in Fig. 5.1 on p. 84. From this it can be seen that the liquid iron–carbon alloy starts to solidify at temperatures between about 1550 °C and 1140 °C, depending on the percentage of carbon. The minimum temperature at which the alloy can remain liquid, 1140 °C, is the eutectic temperature, as we saw in the previous chapter, and it gives the composition of the eutectic alloy as 4.3% carbon–95.7% iron. At any temperature below 1140 °C the iron–carbon alloy must have completely solidified. It will be important to keep this fact in mind when we come to look at Figs. 3.2 and 3.3, because here, at first sight, we seem to have another eutectic for the alloy with 0.83% carbon–99.17% iron. However, what we are seeing is the lower left-hand part of Fig. 5.1; the iron–carbon alloy is solid at every point on Fig. 3.3, and point P is not a eutectic but a *eutectoid* – similar to a eutectic, but relating to the simultaneous formation of two constituents from another *solid* constituent, instead of from a melt.

THE STEEL PORTION OF THE IRON–CARBON EQUILIBRIUM DIAGRAM

When the carbon content of steel exceeds approximately 0.2%, the steel responds to quench-hardening treatments which considerably modify the properties of the grain structure and its tensile strength.

The introduction of carbon converts the single-phase structure of pure iron into a multi-phase structure.

We saw in Chapter 1 that when iron is cooled, its crystal lattice changes from the face-centred cubic lattice above 910°C to the body-centred cubic lattice below 910°C (an example of an allotropic modification). An inverse rate cooling curve plotted for pure iron from above 910°C will indicate a thermal arrest as shown in Fig. 3.1, the arrest being associated with the change in the crystal lattice. The addition of carbon to iron induces a second thermal arrest for all carbon contents except the 0.83% carbon alloy, which still exhibits only a single arrest. Changes in the thermal arrest temperatures for a range of alloys from 0% to 1.5% carbon are shown in Fig. 3.2, the main features of this diagram being:

(i) For carbon contents between 0% and 0.83%, the first thermal arrest becomes lower with increased carbon content.

(ii) The second thermal arrest always occurs at the same temperature (700°C).

(iii) The 0.83% carbon alloy exhibits a single thermal arrest at 700°C.

(iv) For carbon contents greater than 0.83%, two thermal arrests again appear with the second always at 700°C.

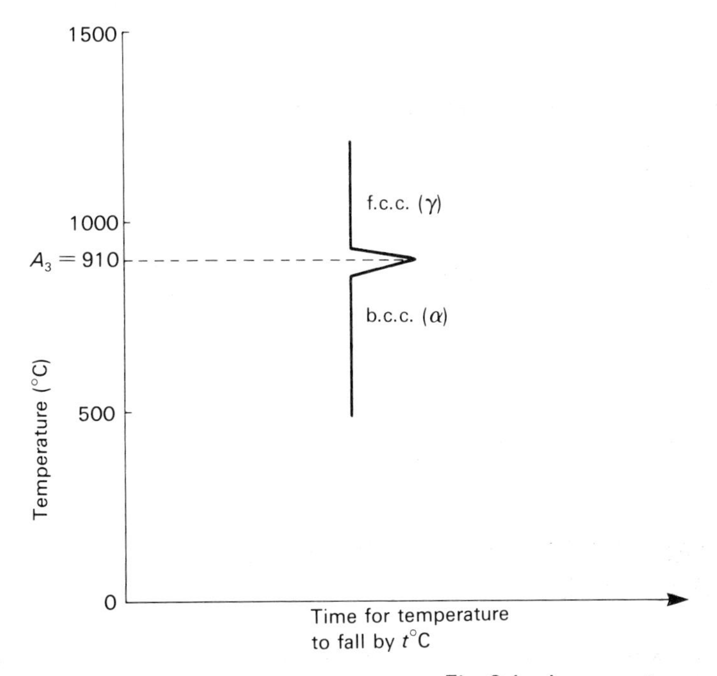

Fig. 3.1 Inverse rate curve for iron

On both sides of the 0.83% value, there is a linear relationship between the temperature of the first thermal arrest and the percentage carbon content.

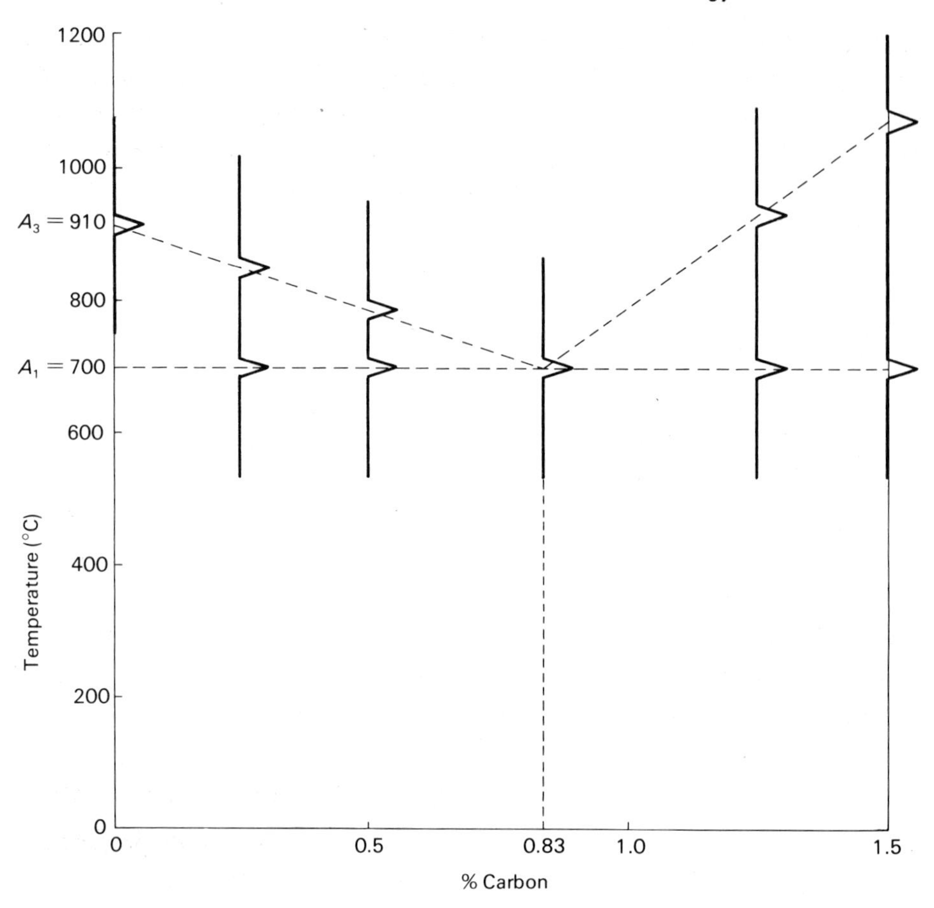

Fig. 3.2 Inverse rate cooling curves for various carbon contents

As the carbon content of steel is increased from 0% to 0.83% the first thermal arrest is found to occupy shorter and shorter periods of time, and the second arrest lasts correspondingly longer. This indicates that an increase in carbon content results in a greater proportion of the alloy sample being associated with this second thermal arrest.

The temperatures of the arrests, known as the *critical temperatures*, are denoted by the letter A, the A_3 temperature being the temperature of the first arrest during cooling, and A_1 being the temperature of the second arrest.*

It is observed that the arrests occur at a slightly higher temperature when the steel is being heated through the critical point, and at a slightly lower

*A further change, A_2, occurs at about 770°C, the *Curie temperature*. This is the temperature at which iron loses its magnetism when it is heated, but this is not important in terms of the structure or properties of plain-carbon steels.

temperature when it is being cooled. To distinguish between these two cases, thermal arrest points which occur during heating are known as A_c points, and those during cooling as A_r points, the notation being derived from the French *chauffage* for heating and *refroidissement* for cooling. The A_c and A_r versions of the A_1 value for steel are approximately 720°C and 695°C respectively.

The iron–carbon equilibrium diagram for steel may be constructed by joining together all the upper critical points and all the lower critical points, as shown in Fig. 3.2.

INTERPRETING THE IRON–CARBON DIAGRAM

Depending on the percentage of carbon and the temperature of the steel, the following products of iron and carbon may appear, as steel is cooled under equilibrium conditions (i.e. slowly):

Austenite, a solution of carbon in face-centred cubic iron (γ-iron). Austenite can only exist above 700°C in slowly cooled steel.

Ferrite, a very dilute solution of carbon in body-centred cubic iron (α-iron) – very soft.

Cementite, a compound of iron and carbon, Fe_3C – very hard and brittle.

Pearlite, a eutectoid mixture of ferrite and cementite, with properties intermediate between the properties of those two constituents.

Above the upper critical temperature (A_3), the steel is uniformly austenite, whatever its percentage of carbon. As the steel is cooled from the upper to the lower critical temperature (A_1), ferrite (if the steel has less than 0.83% carbon) or cementite (if it has more than 0.83%) is precipitated from the austenite. Below the lower critical point, the steel consists of ferrite and pearlite, or cementite and pearlite, respectively.

The equilibrium diagram with these phase fields identified is shown in Fig. 3.3. The phase boundary QP gives the A_3 temperature for any particular carbon content, and is sometimes referred to as the *ferrite solubility boundary*. Taking 0.5% carbon steel as an example, the temperature of the first thermal arrest (the A_3 point), 780°C, will correspond to the commencement of precipitation of ferrite from the austenite. As the temperature falls, the carbon content of the remaining austenite will change down the QP boundary until at 700°C (the A_1 point) the ferrite which has been precipitated will be in equilibrium with austenite containing 0.83% carbon (point P on the diagram). At the temperature of this second thermal arrest, the eutectoid

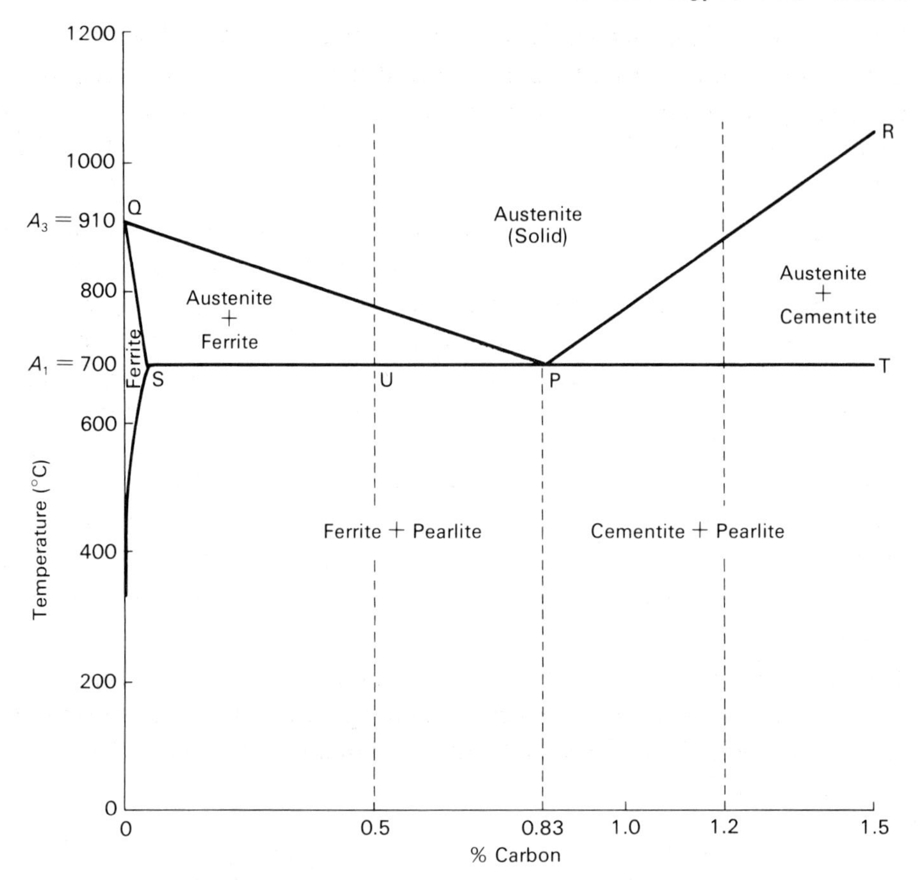

Fig. 3.3 Steel portion of iron–carbon diagram

reaction will occur by alternate precipitation of ferrite and cementite. The final structure will then consist of the primary ferrite precipitated from the austenite between A_3 and A_1, together with the eutectoid mixture of ferrite and cementite known as pearlite. Applying the lever rule we learnt in Chapter 2 gives

$$\text{Ratio}\left(\frac{\text{Ferrite}}{\text{Pearlite}}\right) = \frac{PU}{SU} = \frac{0.33}{0.50}$$

or

$$\text{Percentage of ferrite present} = \frac{0.33}{0.83} \times 100\%$$

$$\approx 40\%$$

Similarly

$$\text{Percentage of pearlite present} = \frac{0.50}{0.83} \times 100\%$$

$$\approx 60\%$$

For slowly cooled plain-carbon steels between 0% and 0.83% carbon, variations in structure will therefore be due only to differences in the proportions of the primary ferrite and the eutectoid pearlite. These will vary from a 100% ferrite structure for 0% carbon steel, to 100% pearlite for 0.83% carbon steel.

The phase boundary RP in Fig. 3.3 is sometimes referred to as the *cementite solubility boundary*, and the temperature indicated by that line for a particular composition is known as the A_{cm} or A_{cem} point. The A_{cm} temperature is the temperature at which cementite will precipitate from the austenite. If we take 1.2% carbon steel as an example, cementite will begin to precipitate at about 900°C. During cooling, the carbon content of the austenite will change down the RP phase boundary until at 700°C the austenite will have a composition represented by P, i.e. 0.83% carbon. The eutectoid reaction will then occur, and the austenite will precipitate alternately ferrite and cementite to form the eutectoid mixture (pearlite). The final structure in this case will therefore be the primary cementite and pearlite.

Typical microstructures for steels with various percentages of carbon are shown in Figs. 3.4 and 3.5. Steels containing less than 0.83% carbon are termed *hypo-eutectoid* steels, and those containing more than 0.83% carbon *hyper-eutectoid.*

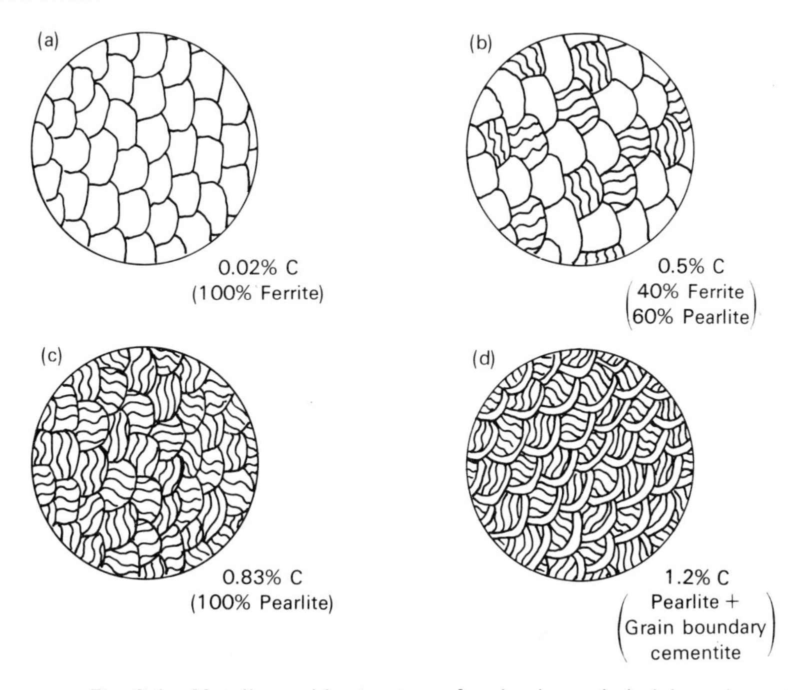

Fig. 3.4 Metallographic structures for slowly cooled plain-carbon steels

Fig. 3.5 Hot-rolled carbon steel – ferrite (white) with unresolved pearlite (black); structure exhibits marked directionality (magnification × 100)

OTHER ELEMENTS PRESENT IN PLAIN-CARBON STEEL

Because of the economics of the steel-making process, a number of residual elements are present in all plain-carbon steels, notably phosphorus, sulphur, manganese and silicon. In some instances, these elements may also be present as deliberate additions to the steel.

Phosphorus

This is normally kept to less than 0.05%, since higher levels tend to result in the precipitation of iron phosphide, Fe_3P, which is brittle, and which segregates in localised areas through the steel.

Sulphur

This too is normally kept to less than 0.05%. It may exist either as ferrous sulphide, FeS, or, if manganese is present in the steel, as manganese sulphide,

MnS. Ferrous sulphide tends to form a brittle intercrystalline film in the grain boundaries, which may give rise to cracking during subsequent working of the steel. Manganese sulphide is a ductile compound that forms within the grains, well dispersed throughout the structure, and it is therefore preferred to ferrous sulphide. Manganese sulphide will form in preference to ferrous sulphide provided that sufficient manganese is present; and for this reason the manganese : sulphur ratio is normally maintained at at least 5 : 1. Sulphur is sometimes added to steel at up to 0.2% to provide well-dispersed MnS inclusions which improve the machinability of the steel by aiding chip formation.

Manganese

This is usually present as a deliberate addition, since it aids pearlite formation and depth of hardening. It does, however, increase the tendency to cracking or distortion during subsequent heat treatment, if present in proportions above about 0.5%. As stated above, it is used to promote the formation of manganese sulphide.

Silicon

This is normally present up to 0.2%, since it is used with manganese as a deoxidiser. Higher percentages than this tend to promote the formation of undesirable graphite from the cementite in the steel.

THE MECHANICAL PROPERTIES OF SLOWLY COOLED PLAIN-CARBON STEEL

We have seen in the previous section that the structure of a slowly cooled plain-carbon steel depends only on the carbon content, and that two different phases or a mixture of these two phases may be present. Ferrite is very soft and ductile; cementite is very hard and brittle; and pearlite, the eutectoid mixture, has intermediate properties. The variation in tensile strength, hardness and percentage elongation (an approximate measure of ductility) with carbon content are shown in Fig. 3.6.

The main features of this diagram are that:

(i) The tensile strength attains a maximum at 0.83% carbon and then diminishes slightly due to the presence of free cementite in the structure, above 0.83% carbon.

(ii) The hardness varies linearly with carbon content.

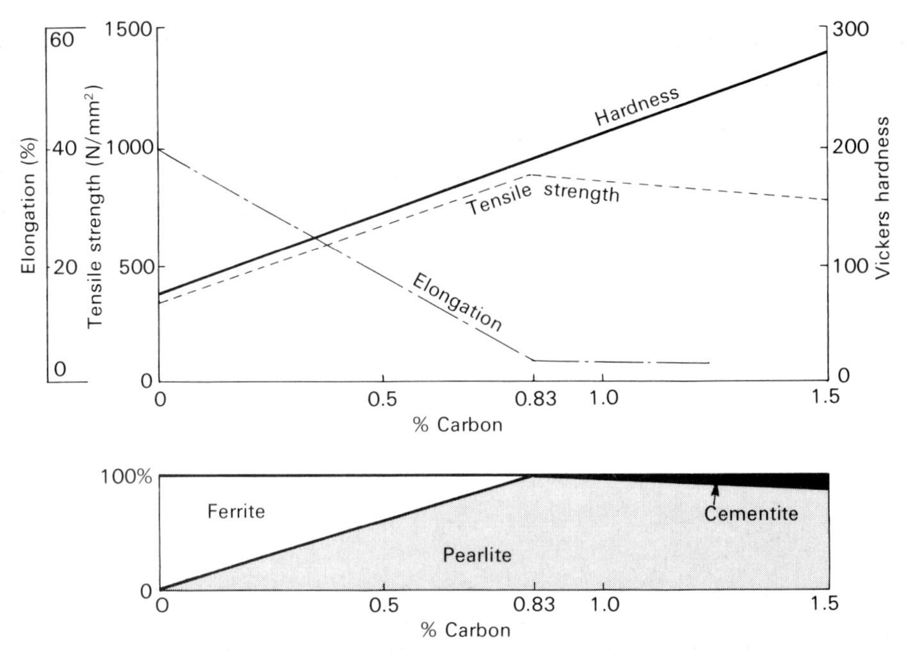

Fig. 3.6 Effect of carbon content on mechanical properties and structure. Annealed condition

(iii) The percentage elongation decreases linearly from 40% for pure iron to about 4% for steel with 0.83% carbon content. At higher carbon contents the percentage elongation value will remain low due to the presence of free cementite in the structure.

Fig. 3.6 is for plain-carbon steels in the annealed condition. We will see, however, in Chapter 4, that major modifications to the properties may be achieved by subsequent heat treatment. While the diagram gives some indication as to the general applications of plain-carbon steels of varying carbon content, specific applications will be considered in Chapter 4.

SUMMARY

- The properties of iron are considerably modified by the addition of carbon. The resulting alloy is called *steel* if the carbon content is not more than about 1.5%, and *cast iron* if it is between about 2% and 4%.

- The equilibrium diagram for the iron–carbon alloys between 0% and 7% C has a eutectic at 4.3% C, which is important for cast irons. There is also a *eutectoid* (similar to a eutectic but relating to the simultaneous formation of two constituents from a *solid* solution) at 0.83% C, and this is important for steels.

- Provided a steel contains not less than 0.2% C, it can be hardened by quenching.

- The iron–carbon equilibrium diagram for the *solid* material is summarised in Fig. 3.7. The A_1, A_3 and A_{cm} temperatures are not exact, because they are higher (A_c) when the steel is being heated, and lower (A_r) when it is being cooled.

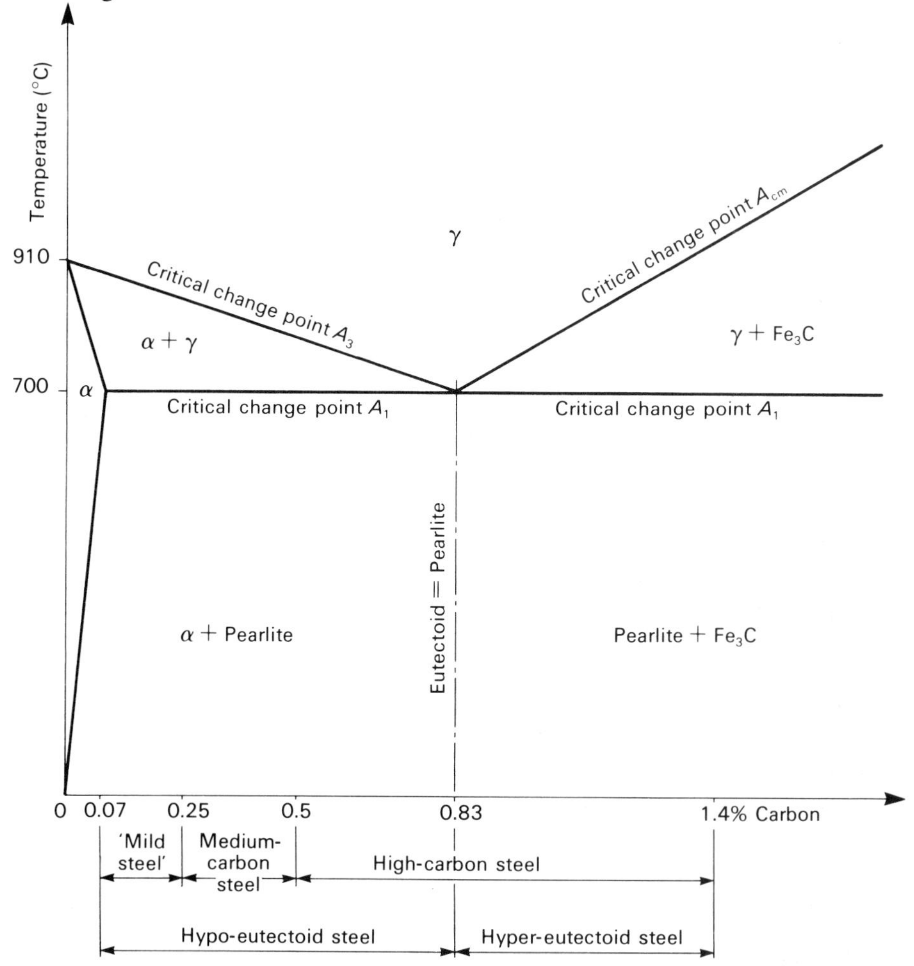

Fig. 3.7 Solid-state equilibrium diagram for steel:

γ = austenite, a solid solution of carbon in face-centred cubic iron

α = ferrite, a very dilute solid solution of carbon in body-centred cubic iron – soft, ductile

Fe_3C = cementite (iron carbide), a hard, brittle, iron-carbon compound

Pearlite, the eutectoid mixture (hence intermediate hardness) of ferrite and cementite in atlernate 'plates', giving a 'pearly' sheen at low magnifications

- Traces of other elements are usually present in steel. The most common are phosphorus, sulphur and silicon.
- *Phosphorus* should be <0.05% or brittle iron phosphide (Fe_3P) precipitate collects in segregations.
- *Sulphur* should be <0.05% unless at least five times as much *manganese* is added. Sulphur without manganese forms ferrous sulphide (FeS), which is brittle and collects in grain boundaries, where it may cause cracking during forming processes. If enough manganese is present, sulphur forms manganese sulphide (MnS), instead, which is ductile and improves machinability – the upper limit for sulphur in this case is 0.2%. Manganese also aids pearlite formation and depth of hardening, but >0.5% may give cracking or distortion during heat treatment.
- *Silicon*: up to 0.2% is used as a deoxidiser, with manganese. More than this may cause formation of undesirable graphite from cementite.
- In annealed steels, tensile strength increases linearly with carbon content up to 0.83% C, and then diminishes slightly as carbon content is increased further. Hardness increases linearly as carbon content is increased from 0% to 1.5% C. Percentage elongation decreases linearly from 40% for pure iron, to 4% at 0.83% C; for greater carbon contents it can be assumed constant at 4%. However, the more carbon in the steel, the more drastically will these properties be modified by heat treatment.

EXERCISE 3

1) Explain the difference between a *eutectic* and a *eutectoid* reaction.

2) Define the following constituents found in plain-carbon steels: ferrite, cementite, pearlite, austenite.

3) (a) Sketch the iron–carbon equilibrium diagram between 0% and 1.2% C.

(b) Sketch an inverse rate cooling curve for a 0.4% C steel, giving the approximate temperatures of any thermal arrests.

(c) Sketch and label the final structure obtained after slow cooling from above the upper critical (A_3) temperature.

4) Assuming the eutectoid reaction occurs at 0.83% carbon, state the phases present and apply the lever rule (Chapter 2) to give the ratio of the phases present for slowly cooled specimens of: (a) a 0.2% carbon steel; (b) a 0.6% carbon steel.

5) Explain the meaning of the A_1, A_3 and A_{cm} critical points.

6) Why is it usual to have sufficient manganese present to ensure the formation of manganese sulphide in preference to ferrous sulphide inclusions in a steel?

7) With the aid of a diagram show the effect of carbon content on the mechanical properties of a plain-carbon steel.

8) Explain the terms *hypo-eutectoid* and *hyper-eutectoid composition.*

9) Sketch and label the metallographic structure of a slowly cooled: (a) 0.8% carbon steel; (b) 1.2% carbon steel.

10) State the carbon content you would select for a plain-carbon steel to be used for the following cases:

(a) a bearing where wear resistance was of prime importance;

(b) an application involving very severe cold forming;

(c) a moderately highly stressed shaft.

4 THE HEAT TREATMENT OF PLAIN-CARBON STEELS

THE DIFFERENT PROCESSES

The four most important heat treatment processes are annealing, normalising, hardening and tempering. The relationship between these treatments, and the appropriate temperatures are shown in Fig. 4.1. Typical structures obtained after the various heat treatments are shown in Fig. 4.2.

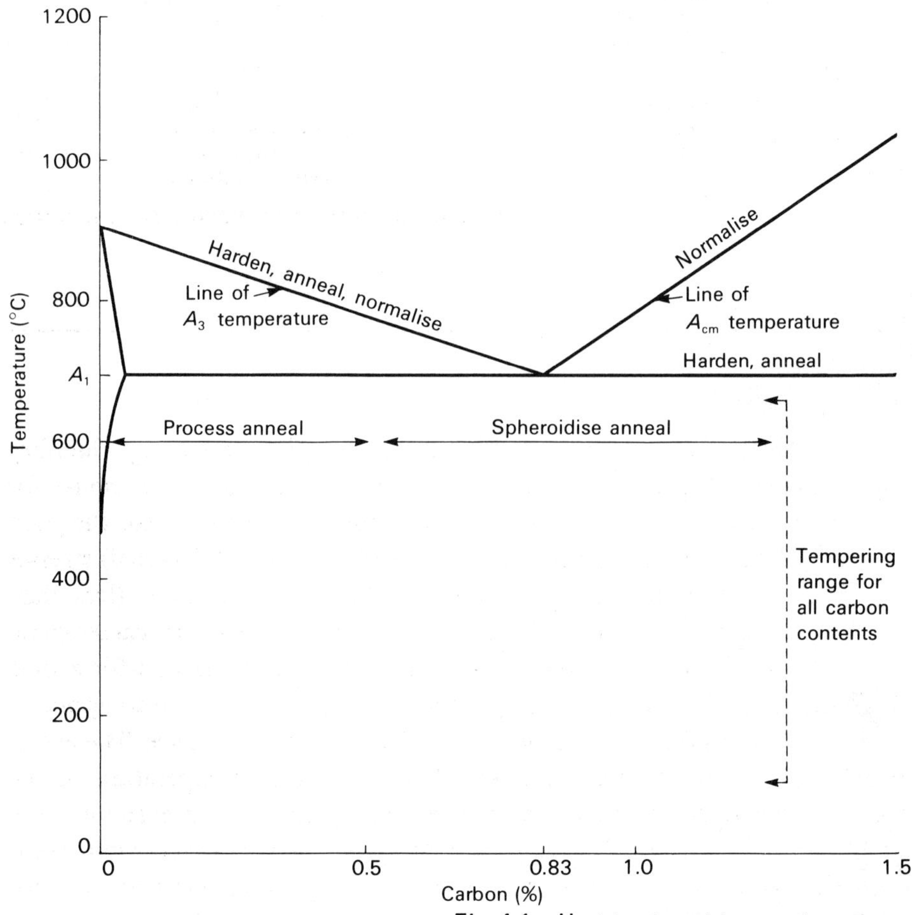

Fig. 4.1 Heat treatment temperature ranges

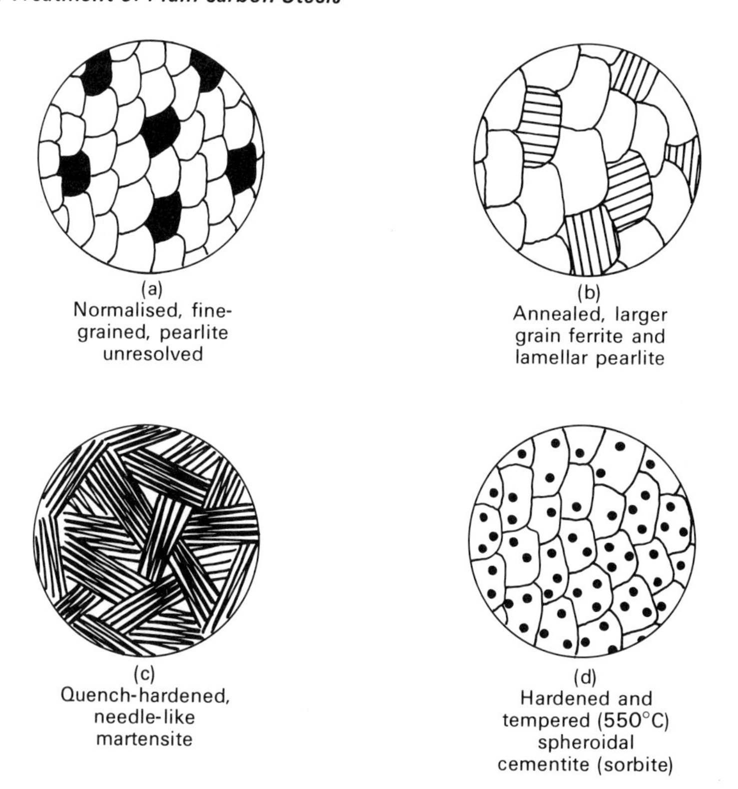

Fig. 4.2 Effect of heat treatment on 0.3% carbon steel

ANNEALING

Full annealing

This heat treatment process produces a steel which is soft, has high ductility and is tough. The process may be carried out for various reasons; for example, to facilitate a severe cold-forming operation, or to make the grain structure of the steel more uniform, or to reduce *residual* (internal) stresses set up by forging or casting operations. For hypo-eutectoid (less than 0.83% C) steels, the component is heated to above the upper critical temperature (A_3), and left in the furnace at this temperature ('soaking') for a time which depends on the mass and cross-sectional thickness of the component. At the end of this time it is allowed to cool slowly in the furnace. The actual annealing temperature used is 30–50°C above the A_3 temperature of the steel. At this temperature the structure will be completely austenite (the equilibrium structure at this temperature). The slow furnace cool will result in an equi-axed ferrite–pearlite structure, free of internal stresses and very ductile. The time for which the steel is 'soaked' at the annealing temperature is fairly critical, since prolonged 'soaking' will give coarse austenitic grains, resulting in a coarse ferrite–pearlite structure with reduced ductility.

Hyper-eutectoid (above 0.83% C) steels are annealed at a temperature just above the lower critical temperature (A_1) since 'soaking' above the A_{cm} phase boundary would result in a coarse structure leading to cracking during subsequent quench-hardening treatment. The mechanical properties of a fully annealed steel will vary with carbon content (see Fig. 3.6, p. 54).

Full annealing of a steel is a costly process because:

(i) it involves heating and 'soaking' the component at a fairly high temperature;
(ii) the slow furnace cool necessary limits the availability of furnaces, and the process is time-consuming.

Process anneal and spheroidising anneal

For economic reasons, low-carbon steels are often *process-annealed* or *subcritical-annealed*. This means heating the steel to just below the A_1 temperature, with the result that the (mainly) ferrite structure recrystallises to a finer grain structure. Also, any residual stresses which may have been present due to cold working are relieved by this treatment.

Hyper-eutectoid (greater than 0.83% C) steels may be softened by a *spheroidising anneal*, by heating to 700°C, just below A_1.* This gives a structure consisting of a ferrite matrix containing cementite in spheroidal (i.e. roughly spherical) form. The material is soft, tough and has good machining properties.

NORMALISING

This means heating the steel to above the A_3 temperature for hypo-eutectoid steels and above A_{cm} for hyper-eutectoid steels, 'soaking' for a time which depends on mass and cross-section thickness, and then allowing it to cool 'normally' in air. The pearlite structure is finer than is obtained after a full anneal and the material is slightly harder and tougher, and has a higher tensile strength than the fully annealed material. Where the slightly decreased ductility and lower degree of stress relief are acceptable the process is preferred, because it is less costly than full annealing.

*The A_1 temperature would be higher than 700°C because the steel is being heated up (see p. 49).

HARDENING

As the term implies, hardening is the form of heat treatment applied to steels to produce the hardest structure obtainable. The treatment consists in heating to just above the A_3 temperature for hypo-eutectoid steels, and above the A_1 temperature for hyper-eutectoid steels, followed by a rapid quench in brine, water or oil. The rapid quench prevents the normal equilibrium conversion of the austenite to ferrite and cementite, so that the austenitic structure is retained below A_1. The retained austenite below A_1 is a non-equilibrium structure, however, and at some temperature between 350°C and 150°C a spontaneous conversion of austenite to a highly strained body-centred lattice occurs, giving a metallographic structure known as *martensite*.

Under the optical microscope, the martensite has a needle-like structure which is very hard and brittle. The hardness of the martensite increases with the carbon content of the steel.

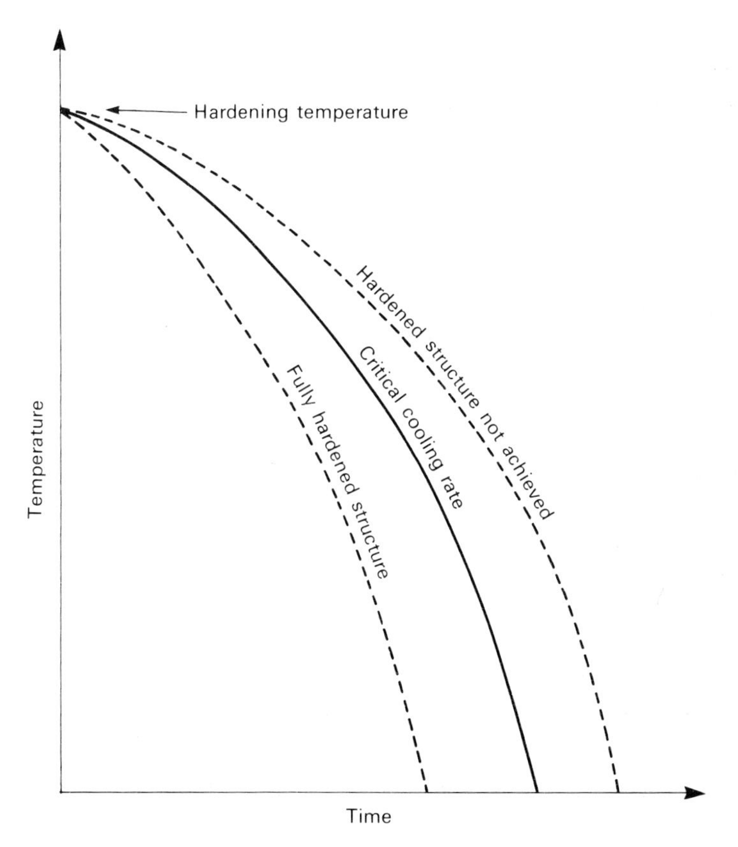

Fig. 4.3 Critical cooling rate

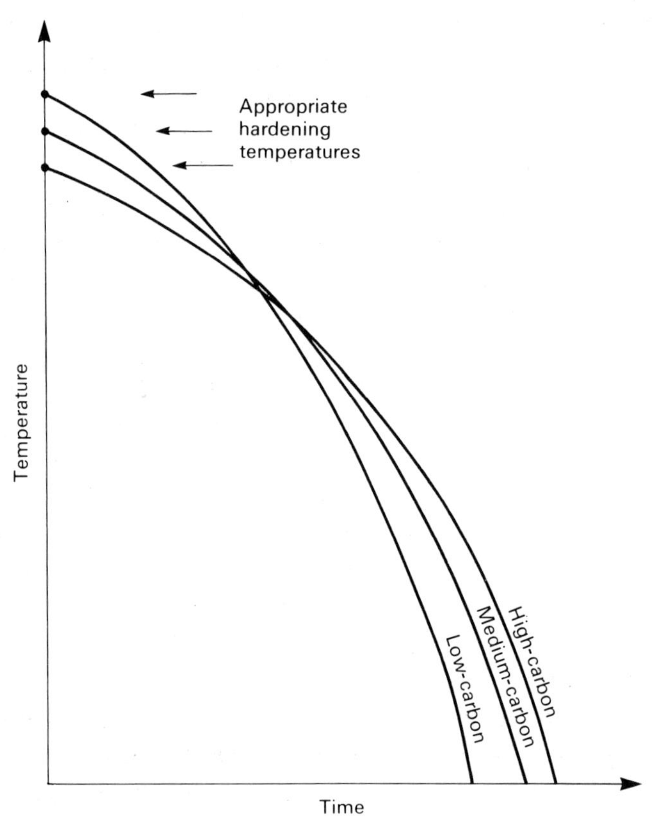

Fig. 4.4 Variation of critical cooling rate with carbon content

In order to suppress the transformation to ferrite and cementite which would occur between A_3 and A_1 during cooling, the cooling rate of the steel must exceed a certain value known as the *critical cooling rate* (Fig. 4.3). The critical cooling rate depends upon the carbon content of the steel – the higher the carbon content, the lower the critical cooling rate (Fig. 4.4). The cooling rate in quenching from the appropriate temperature is a function of:

(i) the mass and shape of the component;
(ii) the ratio of the mass of the component to the volume of the quench medium;
(iii) the characteristics of the quench medium itself.

Quench hardening is liable to produce cracks or distortion in a component, due to the sudden structural transformation to martensite and the thermal stresses set up by the 'shock' cooling. For this reason, we use the least severe quench medium which will enable the critical cooling rate to be achieved. Quenching media are discussed on p. 65.

Material further from the surface of a component will obviously be cooled more slowly by the quench, and if the thickness of cross-section exceeds a certain value, the core of the component will not have a martensitic structure. The term *hardenability* is used to describe the depth of hardening which may be achieved, and, as would be expected from Fig. 4.4, a high-carbon steel will have a better hardenability than a low-carbon steel. One of the main reasons for adding alloying elements to steel is to improve its hardenability. In practice this may mean that oil quenching can be used to achieve full-through hardness of an alloy steel rather than the water quench required for a plain-carbon steel, thus reducing the possibility of distortion or quench cracking. The hardenability of steel is measured by the *Jominy end-quench test*, in which a standard specimen is heated to the normal hardening temperature of the steel and then transferred to a jig by means of which a controlled jet of water is directed against the bottom end of the specimen. When the specimen is cool, hardness measurements are carried out along the length of the specimen, and a curve is plotted of hardness against distance along the specimen (Fig. 4.5). In this diagram, steel A will have a higher martensitic hardness than steel B, which will, however, have the higher hardenability. In practice, the hardenability may be measured by sectioning a component after hardening, and carrying out a hardness scan across the section, as in Fig. 4.6. This once again shows steel A to have the higher martensitic hardness but steel B to have the higher hardenability. In steel specifications it

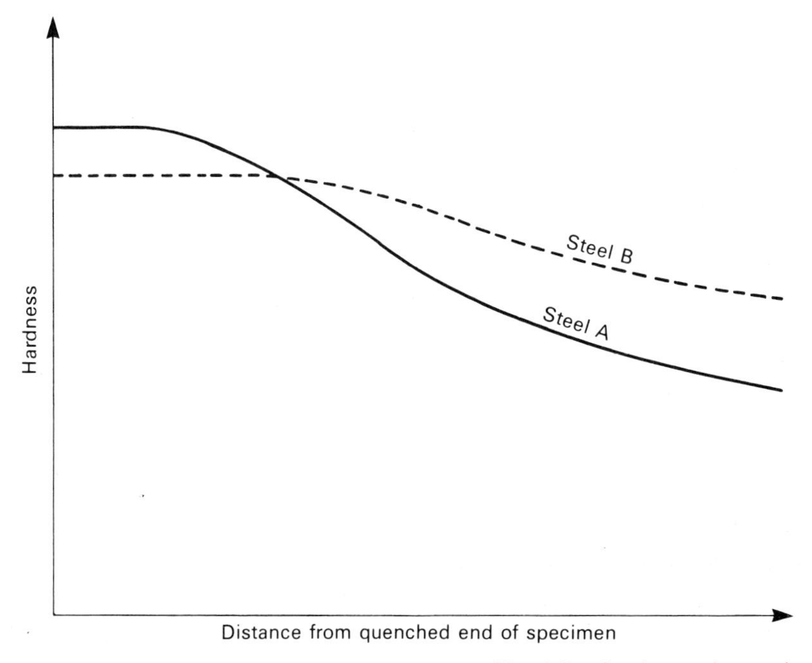

Fig. 4.5 Jominy end-quench values

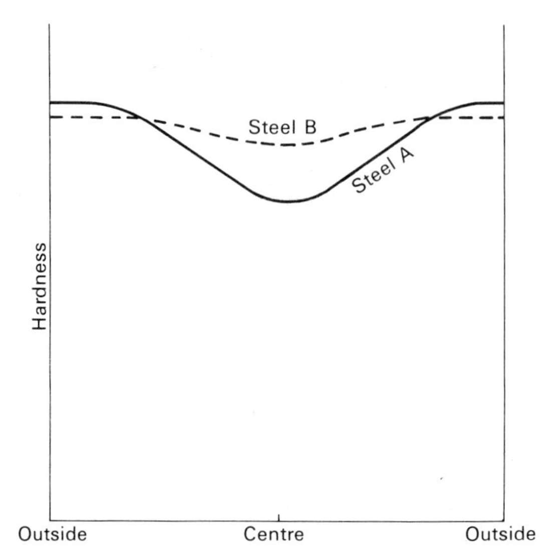

Fig. 4.6 Hardness scan through section of component

is often the practice to quote *ruling sections*, i.e. a maximum section diameter beyond which full martensitic hardening will not be achieved.

THE EFFECTS OF INCORRECT ANNEALING, NORMALISING OR HARDENING TEMPERATURES

'Soaking' a component at too low a temperature will result in failure to achieve the desired metallographic structure, and will give non-uniform mechanical properties. If detected, this may be rectified by reheating to the correct temperature, and then cooling at the appropriate rate. No permanent structural damage will result from this.

In any of the treatments, however, heating to too high a temperature may be serious. Overheating the steel to just above the correct temperature will cause the grain size to increase, with possibly an adverse effect on the mechanical properties. The properties may be restored by reheating to the correct temperature. If excessively high temperatures are used, however, the steel may be 'burnt', resulting in the melting of grain boundary phases, as shown in Fig. 4.7, and this will lead to cracking during subsequent working. The Izod impact test will readily detect a 'burnt' steel, since its impact strength will be drastically reduced. The burnt steel structure cannot be rectified, and in this case the component must be considered as 'scrap'.

Overheating or 'burning' of steels may occur during the heating of a forging prior to the hot working operation. For this reason, on critical forgings,

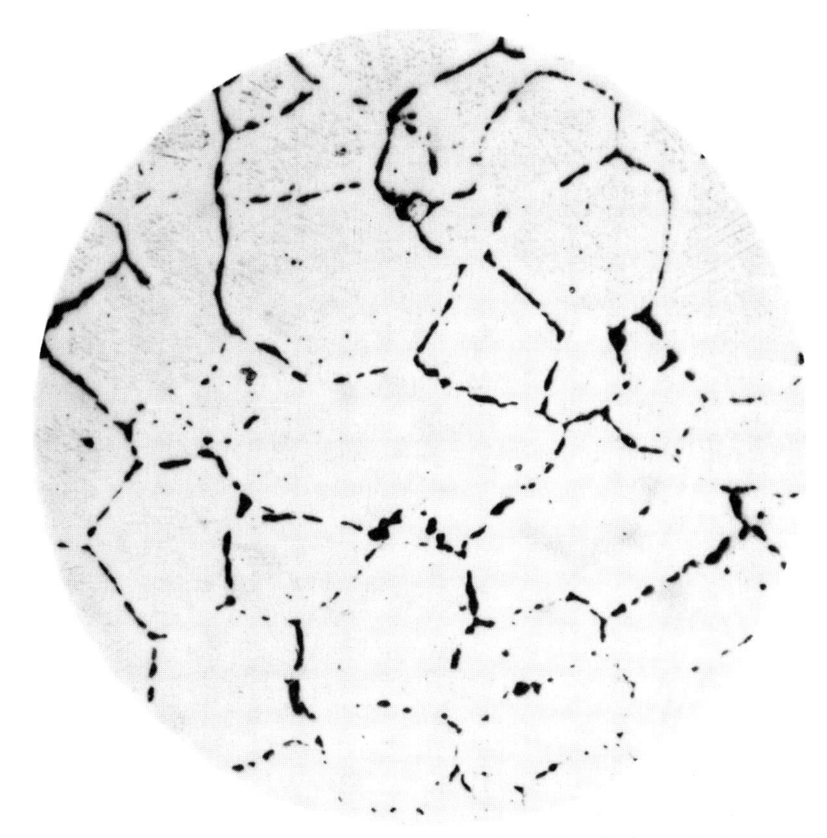

Fig. 4.7 Burnt carbon steel, showing melting of grain boundaries (magnification × 400)

it is sometimes the practice to add an additional piece of material to the thinnest section (the region most likely to suffer overheating) so that, after forging, it can be cut off and tested for overheating or 'burning', by means of either a metallographic examination or an impact test.

QUENCHING MEDIA AND TECHNIQUES

The four main quenching media, indescending order of cooling rate, are brine (a saturated solution of common salt in water), water, oil and air. In industry, however, many other quench media may be used, including solutions of sodium hydroxide and a wide variety of oils with varying characteristics. Because of their higher critical cooling rate, plain-carbon steels usually require a severe quench in brine or water to achieve effective hardening. In the higher-carbon steels, however, if some degree of hardness can be sacrificed to minimise distortion, oil quenching may be used. A number of factors will determine the effectiveness of a particular quench medium, such as its thermal conductivity, its specific heat capacity, and the temperature at which it vaporises. Also the flow characteristics of the medium are important, since they will to some extent determine the rate at which heat is dissipated by convection currents during quenching.

One of the chief disadvantages of brine or water quenching is the formation of steam bubbles around the component. This reduces the cooling rate because of the lower thermal conductivity and lower specific heat capacity of the vapour. This is the reason why the cooling curves of Figs. 4.3 and 4.4 get steeper as the temperature falls: the vapour 'blanket' around the component gets thinner. The higher cooling rate of brine is largely due to deposited salt on the layer of scale which covers the steel. The salt tends to remove the scale during the quenching operation, and since scale has a low thermal conductivity, this gives improved quenching rates.

Oil is a very much less severe quench medium than brine or water. Because of its higher vaporisation temperature, the formation of vapour bubbles around the steel is not so significant. Oil tends to decompose with use, however, which reduced its effectiveness, and thus fairly regular changes are required.

The general principle to be followed in selecting a quench medium is to select the least severe quench medium compatible with achieving the required properties.

Since quenching may result in distortion and cracking, care must be taken to use correct techniques, i.e.:

(i) There should be continuous agitation of the medium to aid uniform heat extraction from the component.
(ii) Long thin sections should be quenched vertically.
(iii) Flat sections and plate or sheets should be quenched edge on.
(iv) In the case of components of varying thickness the thick section should be quenched first.
(v) A high volume ratio of quench medium to volume of component should be maintained.
(vi) The temperature of the quench medium should be maintained within close limits.
(vii) During quenching, the component should not be allowed to come into contact with the sides or bottom of the quench tank.

TEMPERING

The martensitic structure obtained after quench hardening is very hard and brittle, and high residual stresses are left in the material. A further heat treatment known as *tempering* is therefore carried out immediately after hardening. The treatment consists in heating the steel to some temperature below the A_1 temperature, 'soaking' at that temperature, and then, if the steel is a plain-carbon steel, allowing it to cool in air. Some alloy steels may require faster cooling rates than this.

Tempering is necessary for the following reasons:

(i) to reduce the level of residual stress arising from the quenching operation;
(ii) to minimise the distortion which might arise during subsequent working or in service if residual stresses are not removed;
(iii) to modify the mechanical properties and improve the toughness of the steel.

Any structural changes during tempering occur slowly, and therefore it is important that adequate 'soak' time is allowed. If the purpose of the tempering is only to relieve residual stresses and minimise subsequent distortion, temperatures in the range of 100–250°C are used. At these temperatures, no change in the martensitic structure takes place but some redistribution of the internal stresses will be achieved and thus the tendency to cracking will be reduced. No significant change in the mechanical properties will be observed. The use of higher tempering temperatures, however, will result in modification of the martensitic structure, with a corresponding variation in mechanical properties. The general effect of tempering temperature on mechanical properties is summarised in Fig. 4.8.

SURFACE HARDENING OF PLAIN-CARBON STEELS

The surface-hardening techniques applicable to plain-carbon steels are *flame hardening*, *induction hardening* and *carburising*. The first two treatments involve local hardening of a steel which already has sufficient carbon (>0.2% C) to respond to quench hardening. Carburising, however, actually increases the carbon content of the surface layers of a low-carbon steel (mild steel).

Flame hardening

Flame hardening is also known as 'Shorterising', after its inventor. In this process, the surface layer of steel is locally heated, by oxyacetylene torch, to a temperature above the appropriate A_3 temperature, and then quenched by a water spray. It is normally used on medium-carbon steels of about 0.4–0.5% C and the *case depth* (depth of hardened material) obtained is of

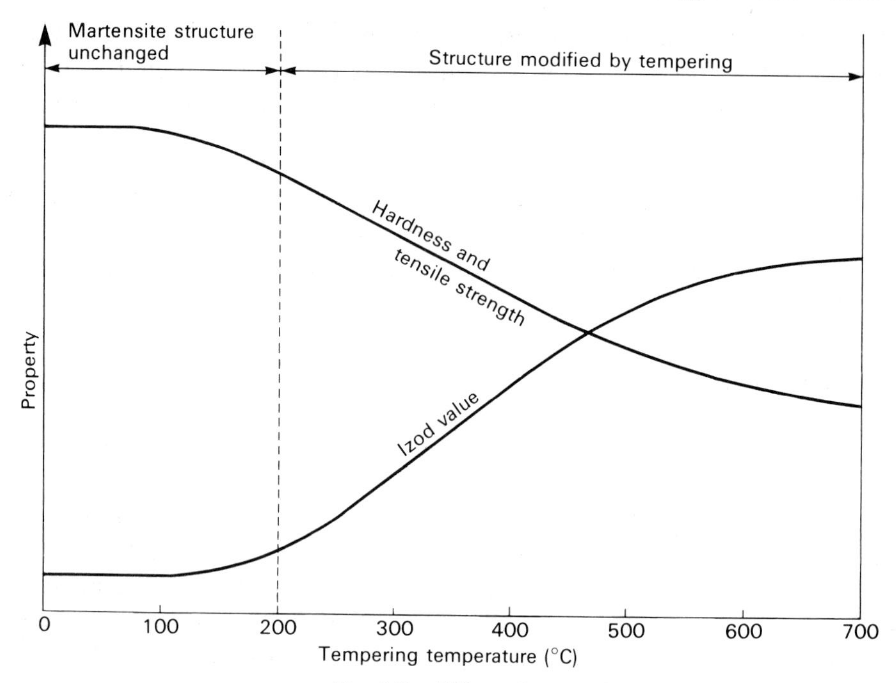

Fig. 4.8 Effect of tempering temperature on properties

the order of 2–3 mm. In its simplest form the process may be carried out by means of a hand-held torch, but the degree of control possible under these conditions is very limited, and it is more usual to apply the process using suitable jigs with either the work moving relative to a series of torches or vice versa. In its most sophisticated form the process is automated, with the flame applied for a controlled time and then replaced by water spray jets for quenching, in a pre-set sequence (Fig. 4.9).

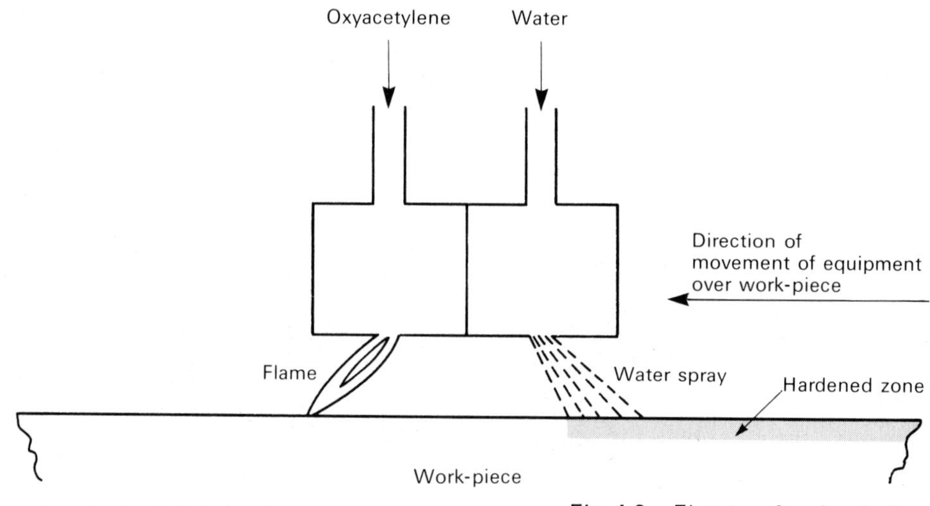

Fig. 4.9 Flame surface hardening

The components should be stress-relieved by a full anneal or normalising before they are flame-hardened. In some cases the process is carried out on components which have been subjected to a full through hardening and tempering.

After flame hardening, a tempering treatment should be carried out, the choice of tempering temperature depending on the final properties required of the case. The process is used to locally case harden items such as large gears, shafts and axles.

Induction hardening

In this process the case of a medium-carbon steel is locally heated above the A_3 temperature by passing it through a coil carrying a high-frequency current (Fig. 4.10). The required temperature is achieved by the heating effect of eddy currents, which are induced in the steel by the high-frequency magnetic field of the coil, and which circulate close to the surface of the material. Before induction hardening, the component should have been either through hardened and tempered, or normalised, to reduce internal stress gradients.

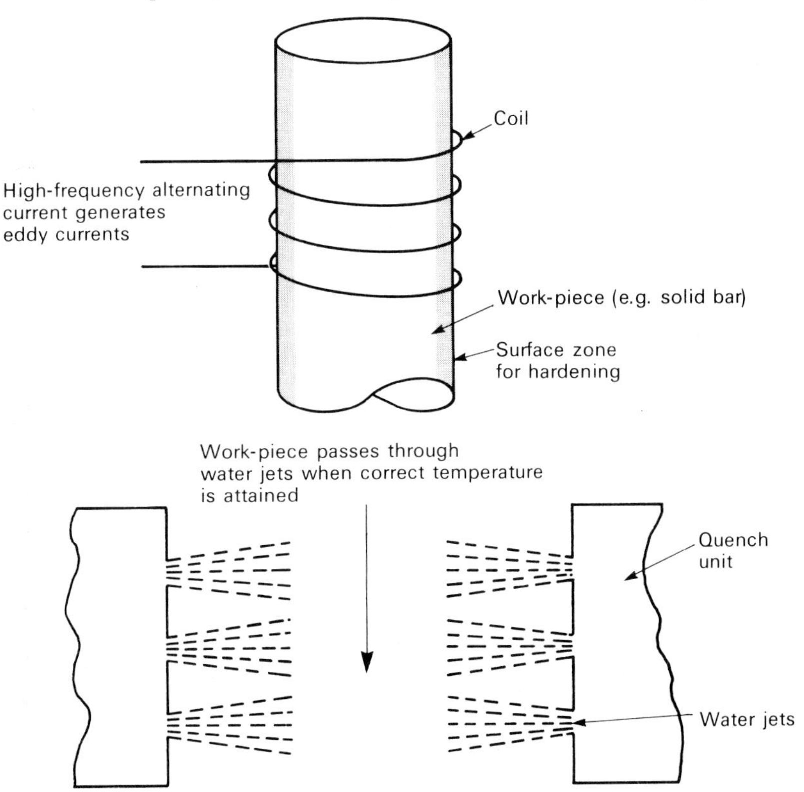

Fig. 4.10 Induction surface hardening

The induction heating machine normally has a quench or spray unit as an integral component of the equipment. Coils may be wound to fit both the shape of the component and the pattern of surface hardening required for a given application. The heating period (normally a few seconds) is automatically time-cycled to suit specific applications. The short period of heating involved in the process reduces the tendency to distortion. A particular form of the process developed for the local hardening of crankshafts is sometimes known as the Tocco process. The process is particularly suitable where production quantities of a component require surface hardening, since high production rates may be achieved. Case depths of a similar order to those obtainable by flame hardening may be obtained (about 3 mm).

In both flame and induction hardening it is important to ensure that there is no decarburisation of the surface of the material, or soft spots in the surface of the component will result. The causes of surface decarburisation will be discussed later in this chapter.

Both processes are also applicable to alloy steels which respond to quench hardening.

Case hardening by carburising

For this process, the component is immersed in a carbon-rich material which may be either solid, liquid or gaseous. Carbon diffuses into the surface of the component to give a case having a carbon content of about 0.83%, which may be subsequently heat-treated to give a hard wear-resistant surface. The process is carried out on low-carbon mild steel components, although a variation in properties may be obtained by using suitable alloy steels.

PACK CARBURISING

The basic process consists in placing the components, usually slightly oversize, in cast iron pots containing the carburising material. This is a mixture of charcoal, which provides the carbon, and barium carbonate, which improves the efficiency of the process. The pots are sealed with fireclay, and the carburising process is carried out by heating the sealed pots in a furnace held at 900–950°C for a time which depends on the depth of case required. During the process, carbon is released by the decomposition of carbon monoxide (CO) according to the reaction

$$2CO = CO_2 + C$$

This carbon diffuses into the outer layers of the component at a rate which depends upon:

(i) the carburising temperature;
(ii) the efficiency of the carburising mixture;
(iii) the composition of the steel.

After carburising, the pots are allowed to cool before the components are removed.

For some applications, there may be a requirement to leave part of the surface uncarburised. This can be done in two alternative ways:

(i) the areas to be left uncarburised are electro-plated with copper, since this acts as a barrier to carbon diffusion; or
(ii) unwanted carburisation is removed by machining – these areas must, of course, have been made sufficiently oversize to permit this, and the machining must be done before quenching hardens the case.

Usually, case depths of between 0.4 mm and 1.5 mm are achieved by carburising. The case has a composition close to the eutectoid composition (0.83% C), while the composition of the core is unaffected.

The prolonged heating at temperatures in the range 900–950 °C causes grain growth in the material of both the core and the case. Because the carbon contents of core and case are so different, heat treatment for grain refinement must be carried out in two stages: the first stage refines the core, the second, the case. (This is eomtimes called a *double refining operation*.)

Core refining. The component is heated to 870 °C, just above the A_3 temperature of the core material (0.1–0.2% C). After 'soaking' at 870 °C the component is quenched in water or oil. This treatment converts the austenitic structure formed at 870 °C into a dispersion of ferrite in martensite. The effect on the *case* is to take any free cementite into solution and the quench will prevent the formation of new free cementite networks and give a coarse martensitic structure. Because of the coarseness of the martensitic case, the second stage of heat treatment, case refining, must follow.

Case refining. The component is heated to 760 °C, just above the critical temperature for a 0.83% C steel, and then quenched in water. This gives a fine martensitic structure, not so brittle as the coarse martensitic structure obtained after core refining. The treatment also modifies the core structure to give a fine ferrite matrix and martensite structure which is tougher and more ductile than that obtained after the core refining treatment.

It is usual to follow these two treatments by a temper at 150–200 °C to relieve quenching stresses.

LIQUID CARBURISING

This consists in heating the component in a salt bath at 900–950°C. The bath contains a mixture of molten salts such as sodium cyanide, barium chloride, sodium carbonate and sodium chloride. After 'soaking' for a time which will depend on the case depth required, the components are usually quenched directly from the salt bath.

Liquid carburising gives shallower case depths. It is therefore often used in tool room work, because it reduces the risk of distortion and enables faster production rates to be achieved. To obtain optimum properties, core refining and case refining heat treaments would be required. In this case, both carburising and some nitriding occur, due to the formation of iron nitride and other nitrides in the surface layer.

The chemical reactions involved in liquid cyanide hardening are probably of the type

$$2NaCN + 2O_2 = Na_2CO_3 + CO + N_2$$

The carburising is done by the CO (carbon monoxide), as indicated below:

$$2CO + 3Fe = Fe_3C + CO_2$$

The nitrogen induces *nitriding*, which is another kind of case hardening:

$$\left.\begin{array}{ll} & N + 2Fe = Fe_2N \\ \text{or} & N + 4Fe = Fe_4N \end{array}\right\} \text{iron nitrides}$$

At the temperature of 900°C the carburising reaction is the more significant.

Cyanide salts are deadly poisons, and the salt bath should be fitted with efficient fume extraction equipment. Care must also be taken to avoid contact with the salts, as cyanide may enter the blood stream through an open cut. For the same reason, no food or drink should be consumed in the heat treatment shop.

If cyanide has been swallowed by an operator it is absolutely essential that an antidote be swallowed *immediately*. For this reason an antidote should be readily available in the heat treatment shop. A doctor should be summoned *immediately*.

A recommended antidote consists in a mixture made up of

(i) 158 g BP ferrous sulphate ($FeSO_4 \cdot 7H_2O$) + 3 g citric acid crystals dissolved in 1 litre of distilled water;

(ii) 60 g of anhydrous sodium carbonate (Na_2CO_3) dissolved in 1 litre of distilled water.

The antidote dose consists of a tumbler made up of equal quantities of (i) and (ii). If the patient is unconscious, artificial respiration should be applied and oxygen inhaled. A capsule of amyl nitrite may be held for inhalation by the patient.

Expert opinion should be sought regarding the formulation of the antidote and the necessity for periodic replacement, since the ingredients may deteriorate with time, particularly in bright sunlight.

GAS CARBURISING

This process is carried out by heating the components at 900–920°C in a carbon-rich atmosphere such as propane. The advantage of gas carburising is that the case depth and the quality of the case may be more closely controlled than in either of the other two processes. Specialised and costly equipment is required.

HEAT TREATMENT EQUIPMENT

There are three main types of equipment for heating steel: (i) non-muffle-type furnaces; (ii) muffle-type furnaces; (iii) salt baths. In non-muffle-type furnaces the steel is often in contact with furnace gases, whereas in muffle furnaces the steel is segregated from such contact, as shown in Fig. 4.11. In special cases the atmosphere inside a muffle furnace may be controlled, to reduce surface decarburisation, which is caused by contact with air or

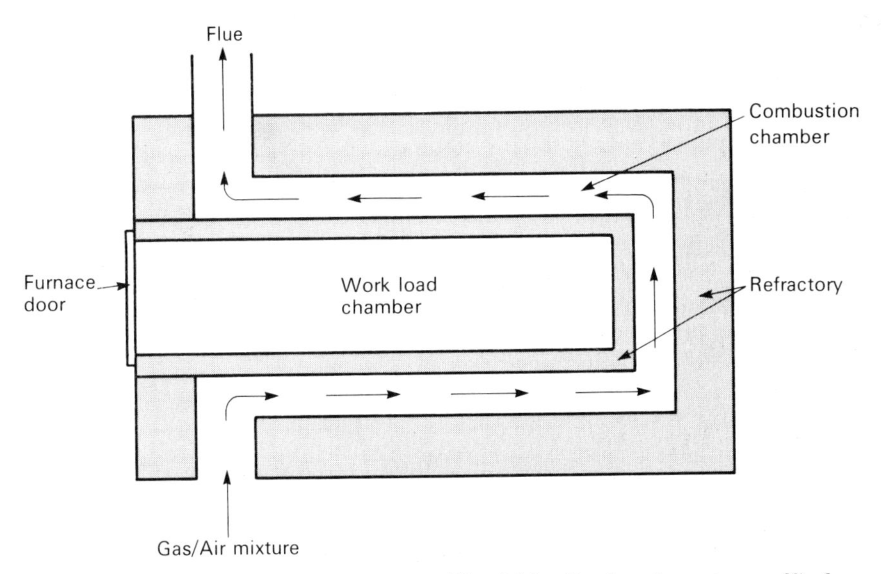

Fig. 4.11 Section through a muffle furnace

furnace gases. The atmosphere in a controlled atmosphere furnace is therefore a non-oxidising (*reducing*) gas such as 'cracked' ammonia which essentially provides an atmosphere consisting of a mixture of hydrogen and nitrogen. Generally, any steel component heated in a furnace without atmosphere control will suffer surface decarburisation, and at higher temperatures (above 900 °C) surface scaling. The use of controlled atmosphere muffle furnaces or vacuum furnaces greatly increases costs, and often salt baths provide a more economic alternative.

Salt-bath-type heat treatment equipment consists essentially of a tank or pot containing a mixture of salts which melt at specific temperatures and thus provide a liquid medium for the heating of steel components. The choice of salts used depends primarily on the temperature of operation, nitrates and nitrites being used for low temperatures, and chlorides and carbonates for the higher temperatures. The liquid carburising process is a special case where cyanide salts are used. Because the molten salt keeps air away from the component, no surface decarburisation or scaling occurs during the heating process.

It is important that the temperature of furnaces and salt baths should be accurately controlled and recorded. Ideally, the equipment should provide a continuous chart read-out of temperature against time, so that a permanent record is available.

Since many heat treatment processes involve quenching in brine, water or oil, quench tanks are an essential part of the heat treatment shop. The main features of this equipment are that it should provide sufficient volume of quench medium with suitable means of agitation, and, in some cases, temperature control. This is sometimes achieved by surrounding the quench tank with a second tank through which cooling water circulates.

Safety precautions

These may be listed as follows:

(i) Suitable temperature control and recording equipment should be provided.
(ii) Heat treatment equipment should be separate from other processes.
(iii) Access to heat treatment equipment must be restricted to trained and knowledgeable personnel.
(iv) Efficient mechanical handling equipment must be available, especially for heavy components.
(v) Efficient ventilation and fume extraction equipment must be provided, since toxic fumes may be present.

(vi) Non-recommended mixtures of salts should be avoided – for instance, mixtures of nitrates and cyanide salts may become explosive.

(vii) Where components are quenched in oil, an air-excluding cover for the quench tank must be available, to extinguish any outbreak of fire which may occur.

(viii) Food or drink must not be consumed in the heat treatment area.

(ix) Any open wounds or sores must be kept covered.

(x) Clear warning signs must be displayed, particularly where air cooling is being used, to avoid accidental handling of hot equipment or components.

(xi) Specialised fire-fighting equipment should be readily available.

(xii) Protective clothing should be worn by operators, e.g. gloves, aprons, masks.

(xiii) A first aid kit should be readily available. Such a kit should contain a cyanide antidote if cyanide salts are being used (see p. 72).

PROPERTIES AND APPLICATIONS OF PLAIN-CARBON STEELS

Plain-carbon steels may be broadly divided into three categories: low-carbon (0.07–0.25% C), medium-carbon (0.30–0.50% C) and high-carbon (0.60–0.80% C). While, within these groups, a fairly wide range of properties may be obtained, *typical* mechanical properties and applications are given in Table 4.1.

Alloy steels are more expensive than plain-carbon steels, and should therefore be avoided as far as possible, but, for some applications, the addition of suitable alloying elements such as nickel, chromium or molybdenum may be necessary to obtain specific properties. In selecting a steel for a particular application, additional factors such as availability, service environment and manufacturing methods may have to be considered, and the designer usually has to compromise in making his final choice of material.

Strictly speaking, the term 'plain-carbon steel' is misleading, since all such steels contain manganese as an alloying element. This gives improved mechanical properties, increases the hardenability and ensures the formation of manganese sulphide rather than the brittle ferrous sulphide within the structure.

TABLE 4.1 *Properties and applications of plain-carbon steels*

Composition		Typical properties				
Range	% carbon	Yield or 0.1% PS† (N/mm^2)	UTS† (N/mm^2)	Elongation (%)	IZOD (J)	Condition
Low-carbon	0.07–0.15	240	380	35	84	As-rolled
		385	415	20	42	Cold-drawn
	0.15–0.25	290	490	33	78	As-rolled
		365	520	35	110	Normalised
		280	515	32	76	Annealed
		570	670	12	30	Cold-drawn
Medium-carbon	0.30–0.50	350	520	30	46	Annealed
		410	610	25	50	As-rolled
		610	920	15	25	*H850°C; WQ, T 450°C
		525	800	20	40	H850°C; WQ, T 550°C
		490	720	24	46	H850°C; WQ, T 650°C
High-carbon	0.60–0.80	370	650	23	17	Annealed
		415	725	20	32	As-rolled
		600	910	20	29	*H840°C; OQ, T 550°C
		550	840	23	31	H840°C; OQ, T 600°C
		510	770	25	45	H840°C; OQ, T 650°C
Hyper-eutectoid	0.90–1.40	600	870	6	3	As-rolled

†PS = proof stress
UTS = ultimate tensile stress

*H = hardened
OQ = oil-quenched
WQ = water-quenched
T = tempered

TABLE 4.1 *continued*

	Comment	Typical applications
	Suitable for welding, cold forming, carburising. Does not respond to quench hardening. Properties depend on degree of cold work.	Rivets, rod, nails, wire, cold-rolled sheet, car body pressings, tubes.
	Does not significantly respond to quench hardening. Suitable for more moderate cold working. May be carburised. Used for hot pressing.	Boiler plate, forgings, steel sections, low-stressed shafts.
	Moderate response to quench hardening. Used where higher properties are required. May be flame- and induction-hardened. Not normally suitable for welding or carburising.	Shafts, higher strength tubes, crankshafts, connecting rods, forgings where moderate strength is required.
	Significant response to quench hardening. Higher hardenability steel. Not suitable for welding, cold forming or carburising.	Springs, saws, forging dies, wire ropes, crankshafts, small gears.
	Used in hardened and tempered condition (tempered to about 200 °C) where maximum hardness rather than tensile strength is required.	Razors, drills, turning tools, shear blades, punches.

SUMMARY

- *Annealing* makes steel soft, tough and ductile, but takes up costly furnace time. It may be carried out to facilitate cold forming, to give uniform grain structure, or to relieve internal stresses. Steels with less than 0.83% C are 'soaked' at 30–50 °C above the A_3 temperature, then slowly cooled in the furnace. This gives an equi-axed ferrite–pearlite structure. Prolonged soaking may reduce ductility due to coarsening of the grain structure. Steels with more than 0.83% C are heated to just below A_1 temperature, to avoid coarsening the grain structure, which might induce cracks if the steel is quenched later.

- *Normalising* is cheaper than annealing because components spend less time in the furnace. This is because after soaking at its A_3 or A_{cm} temperature, the steel is left to cool 'normally' in the air. Normalising gives less ductility and stress relief than annealing and the material is slightly harder and tougher but the pearlite structure is finer.

- *Hardening*. The steel is heated to just above its A_3 temperature (or A_1 if $>$ 0.83% C), then cooled at a rate exceeding its critical cooling rate – usually by *quenching* in oil, water or brine. This converts the austenite to *martensite*, a hard brittle needle-like structure with a strained body-centred cubic lattice. If the critical cooling rate is not attained, the austenite forms ferrite and cementite in the usual way. The critical cooling rate depends on the quench medium, on the mass and shape of the component to be quenched, and on the ratio of the masses of component and quench medium. Severe quenching may cause cracking or distortion of the component, so steel is sometimes alloyed with elements which improve its *hardenability*, enabling a less severe quench medium to be used. Hardenability is determined by a *Jominy test*, in which a specimen is quenched on one end face, hardness measurements then being taken along its length. *Ruling sections* in steel specifications state the maximum diameter which can be hardened right through.

- *Overheating*. If steel is accidentally heated to a temperature just above the correct temperature for annealing, normalising or hardening, it causes growth in grain size. This can be put right by reheating to the correct temperature.

  Heating too far above the correct temperature, however, causes steel to be 'burnt': grain boundary phases melt and the steel becomes brittle and unreliable. There is no remedy; the steel must be scrapped.

- *Tempering* is carried out after quenching, to relieve internal stresses caused by the quench, and to make the steel tougher (less brittle) by moderating the martensitic structure, though this causes some loss of hardness and tensile strength. The component is 'soaked' at the tempering temperature, then allowed to cool in air. Tempering temperatures up to 200°C are for stress relief only; higher temperatures progressively increase Izod value and reduce hardness. The maximum temperature for tempering should not exceed A_1.

- *Surface hardening* may be by *flame hardening* or *induction hardening* if the steel has at least 0.2% C; otherwise the steel must be carburised. In the first two processes the surface of the steel is heated to above the A_3 temperature by gas jets or eddy currents, respectively, and immediately quenched by water jets. To relieve internal stresses, the material should already be in the annealed or normalised state if the properties of the core are not important, or in the hardened and tempered state if they are. This is usually ensured by buying the material in the appropriate state before components are made. After surface hardening, a tempering treatment should be carried out.

- *Carburising* means diffusing additional carbon into the surface layer of material (the *case*). One method is *pack carburising* in which the components are packed in carbon-rich material (charcoal and barium carbonate) in pots which are kept at 900–950°C until the case is deep enough. This also causes grain growth in the core and case of the steel, so after pack carburising the core must be refined by 'soaking' at just above the A_3 temperature of the core material, and then quenching. Then the case must be refined by reheating to 760°C and again quenching. Finally it is advisable to relieve quenching stresses by tempering at 150–200°C.

  A second method of carburising is *liquid carburising* in which the components are 'soaked' at 900–950°C in molten salts (chiefly cyanides), to supply the carbon, and quenched from the salt bath. This gives shallower case depths, but faster production rates. Core and case refining are still advisable.

  A third method is *gas carburising* in which the carbon diffuses in from a gas such as propane, at 900–920°C.

  To limit carburised areas, components can be masked, or the 'case' can be machined away locally.

- *Heating equipment* for heat treatment processes is either of furnace or salt bath type. In the simple furnace, components are in contact with the air and also, possibly, with hot furnace gases. They therefore tend to lose carbon (*decarburise*) at their surfaces, and also to become covered in scale, because they are being heated in an *oxidising* atmosphere which burns off carbon and oxidises iron. *Muffle furnaces* avoid this by heating components in an enclosed compartment, sealed from the air and from furnace gases. The components can be heated in a *reducing* atmosphere; e.g. a mixture of hydrogen and nitrogen, which *reduces* any oxygen present to water vapour, thus preventing decarburisation and scale.

  A cheaper alternative is the molten salt bath, in which the molten salts both transfer heat to the component and keep the air away from it.

- *Liquids used in quenching* are chiefly brine, water and oil. Because quenching may cause cracking and distortion and leave internal stresses, the quenching liquid should be no more severe than is necessary to achieve the required cooling rate. Brine and water are severe quenching liquids, brine especially so, since, in evaporating it leaves salt on the component which removes any scale, giving closer contact with the steel. Also, brine has a higher boiling point than water, and it is the formation of vapour bubbles on the component which limits the effectiveness of any quenching liquid. Although oil has a much higher boiling point, its thermal conductivity and specific heat capacity are less and its viscosity is greater, so it is a much less severe quench medium.

  Oil tends to decompose with use, so it needs regular replacement, which is expensive.

- *Quenching procedures.* Components should be quenched thick end first; thin sections and plates should be quenched vertically, and components should be kept clear of tank surfaces. A large volume of quench liquid with continuous agitation and provision for its cooling may be necessary.

- *Safety precautions.* Because cyanides are deadly poisons, heat treatment should be confined to a separate workshop in which food and drink are banned, all cuts and sores must be covered by dressings, and access is limited to trained personnel. The workshop should have efficient ventilation, fire-fighting, mechanical handling and temperature control equipment. Signs warning of hot equipment or components should be used, and deviations from standard operating procedures must be prohibited to prevent possible poisonous or explosive effects.

- *Applications of steels*

Mild steel up to 0.15% C:	General fabrication. Welds, cold forms, carburises, but is non-quench-hardenable.
Mild steel; 0.15–0.25% C:	Stronger, less ductile. Some cold forming, hot presses, carburises, non-quench-hardenable.
Medium-carbon steel; 0.25–0.5% C:	Moderate quench hardening. Can be flame- or induction-hardened. For shafts, etc., where higher strength is required.
High-carbon steel; 0.5–0.83% C:	High hardenability, high strength, for springs, saws, dies, etc.
Hyper-eutectoid steel: 0.83–1.4% C:	For maximum hardness and sharpness in cutting tools.

EXERCISE 4

1) Describe briefly the full annealing process as applied to a 0.5% carbon steel.

2) State the advantages of the normalising process compared with the full-anneal heat treatment for a carbon steel.

3) With the aid of a diagram explain the term *critical cooling rate* for a carbon steel.

4) How may the hardenability of a steel be assessed?

5) A 0.6% carbon steel is: (a) furnace-cooled from 850°C; (b) water-quenched from 850°C.

In both cases sketch and briefly describe the metallographic structure obtained.

6) Explain the meaning of the terms: (a) an overheated steel; (b) a burnt steel.

7) State four factors which should be considered when quenching from the hardening temperature, in order that the tendency to cracking or distortion may be minimised.

8) With the aid of a suitable diagram show how the tensile strength of a carbon steel is affected by increased tempering temperature.

9) Briefly describe the *pack carburising* method for surface hardening a low-carbon steel.

10) Sketch and label the metallographic structure of (a) the case, and (b) the core, of a pack-carburised and heat-treated low-carbon steel component.

11) Write an account of the safety precautions which should be considered in a heat treatment shop.

12) Explain, with the aid of a diagram, the difference between muffle-type and non-muffle-type heat treatment furnaces.

13) What are the advantages of heat treating steel components in a salt bath as opposed to a normal muffle furnace?

14) (a) What should be the carbon content of a steel to be used as the material of a spring?

(b) What heat treatment should be given to this steel in the manufacture of the spring?

15) Briefly describe the induction-hardening method for surface hardening a steel component.

5 CAST IRONS

THE COMPOSITION OF CAST IRONS

Cast irons are alloys of iron and carbon in which the carbon content is normally between 2% and 4%. Other elements such as silicon, manganese, sulphur and phosphorus are present, and, in the case of alloy cast irons, nickel and chromium are often added in significant quantities. Cast irons have certain specific advantages, and find wide applications where very high tensile strengths are not required. The main advantages are:

(i) comparatively low cost;
(ii) good casting properties, enabling both small and very large castings to be produced;
(iii) a reasonably wide range of properties may be obtained by variations in composition, cooling rates and heat treatment.

The structures obtained after the solidification of a cast iron vary widely, as can be seen from Fig. 5.1. An important feature of this diagram is the presence of a eutectic reaction at 4.3% carbon. This value, 4.3%, becomes modified in the presence of silicon or phosphorus or both in the iron. To allow for the effect of this on the proportions of the equilibrium diagram either side of the eutectic, a *carbon equivalent value* is often used when reading from the standard equilibrium diagram of Fig. 5.1. This is calculated as

$$\text{Carbon equivalent value} = \%\,\text{Carbon} + \frac{\%\,\text{Silicon} + \%\,\text{Phosphorus}}{3}$$

Thus a cast iron having the composition 2.5% carbon, 1.0% silicon and 0.5% phosphorus will have

$$\begin{aligned}\text{Carbon equivalent value} &= 2.5 + \frac{1.0 + 0.5}{3}\\ &= 2.5 + 0.5\\ &= 3.0\%\end{aligned}$$

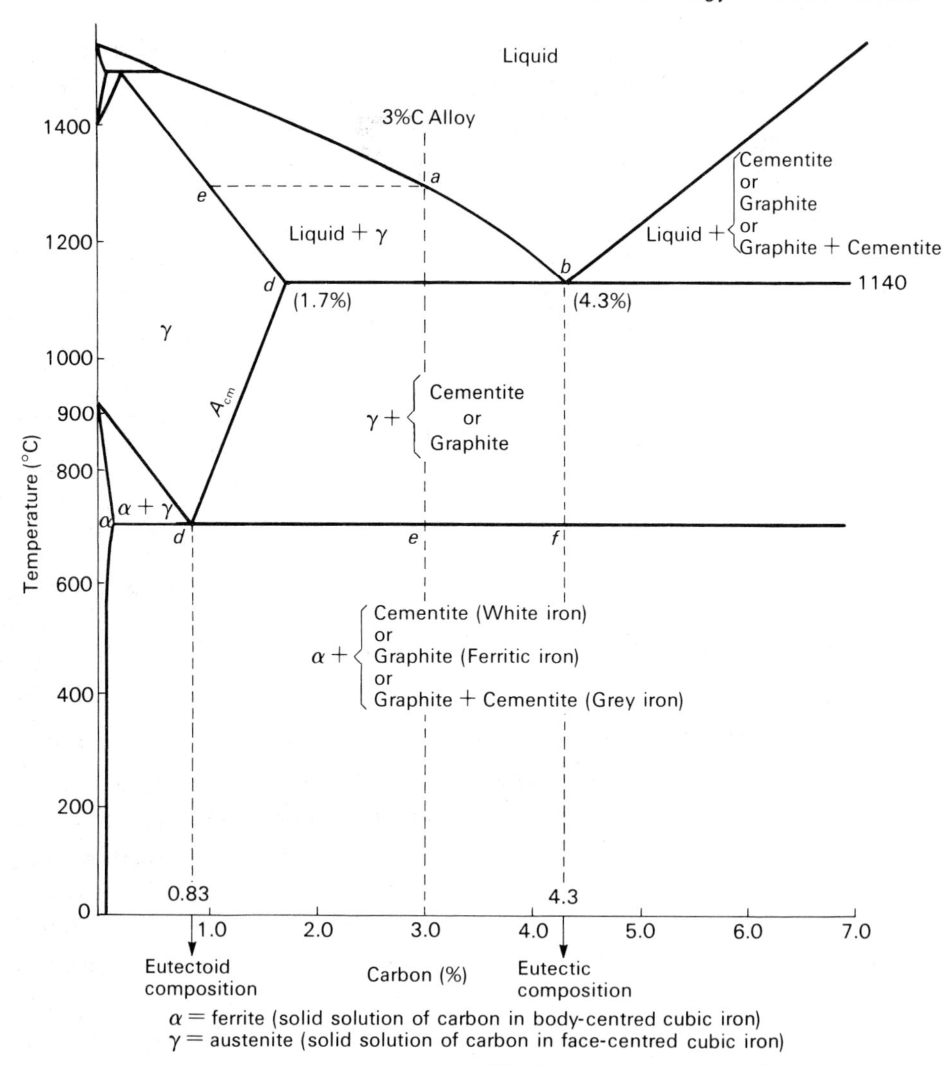

Fig. 5.1 Iron–carbon equilibrium diagram

This cast iron would therefore be treated as if it were a simple 3% carbon cast iron, when reading from Fig. 5.1.

Usually only hypo-eutectic alloys (less than carbon equivalent of 4.3%) are used commercially, since hyper-eutectic alloys are very brittle.

Silicon, which may be present at up to 3%, promotes the formation of graphite by breaking down the cementite present into iron and graphite:

$$Fe_3C = 3Fe + C \text{ (graphite)}$$

Effectively this means that silicon will tend to aid the formation of a grey iron, so called from the appearance of the fracture surface.

Phosphorus, when present, produces an iron–iron-phosphide eutectic which has a low melting point and which therefore improves the casting properties. The phosphorus content is usually kept down to a low level, however, since the eutectic also has the effect of making the iron more brittle.

Sulphur tends to stabilise cementite, preventing its breakdown into iron and graphite, and thus its effect is the opposite of silicon's. Normally the sulphur content is balanced by a similar proportion of manganese, so as to promote the formation of manganese sulphide, rather than iron sulphide which tends to form brittle grain boundary films.

If more manganese is present than is required to take up all the sulphur as manganese sulphide, the excess manganese tends to increase the tensile strength of the iron.

THE SOLIDIFICATION OF CAST IRON

In Fig. 5.1, an iron with 3% carbon is taken as an example. Solidification will commence at 1300°C when initial dendrites of austenite containing approximately 1.0% carbon (*e* on diagram) will solidify from the melt. As cooling proceeds, the composition of the remaining liquid will change down the phase boundary *ab* while the composition of the austenite changes down the solidus, *ed*, shown on the diagram. At a temperature of 1140°C, austenite containing 1.7% carbon (*d* on diagram) will be in equilibrium with liquid solution of composition 4.3% carbon (*b* on diagram). At this temperature the eutectic reaction takes place and the liquid solidifies as a mixture of austenite (1.7% carbon) and cementite or graphite. On further cooling, the solubility of carbon in austenite changes according to the A_{cm} phase boundary, and this results in the further precipitation of either cementite or graphite until the eutectoid composition of 0.83% carbon is attained at temperature 700°C. The eutectoid reaction results in the breakdown of the remaining austenite, of composition 0.83% carbon, to the mixture of cementite and ferrite known as pearlite.

These reactions are summarised below:

Commencement of solidification at 1300°C:

dendrites of austenite containing 1.0% carbon + liquid solution.

At eutectic temperature 1140°C:

austenite containing 1.7% carbon + liquid solution containing 4.3% carbon.

At eutectic temperature 1140°C after the reaction:

austenite (1.7% carbon) + eutectic mixture of austenite + either cementite or graphite.

After eutectoid reaction at 700°C:

conversion of austenite to pearlite.

The final structure is therefore *pearlite + either cementite or graphite.*

If the final structure is pearlite + cementite, usually attained after fast cooling, the cast iron is known as a *white iron* (Fig. 5.2(a)), but if the final structure is pearlite + graphite this is known as a *grey iron* (Fig. 5.2(b)). Very slow cooling may result in excessive precipitation of graphite along A_{cm} to give a final structure consisting of ferrite and graphite, and this is known as a *ferritic iron* (Fig. 5.2(c)). The presence of graphite flakes in the structure of a cast iron will generally result in a low tensile strength and low impact resistance. This is due to the fact that the graphite flakes (*kish graphite*) will act as stress concentration centres, where cracks will readily form during loading, causing premature failure. For this reason, the composition of grey

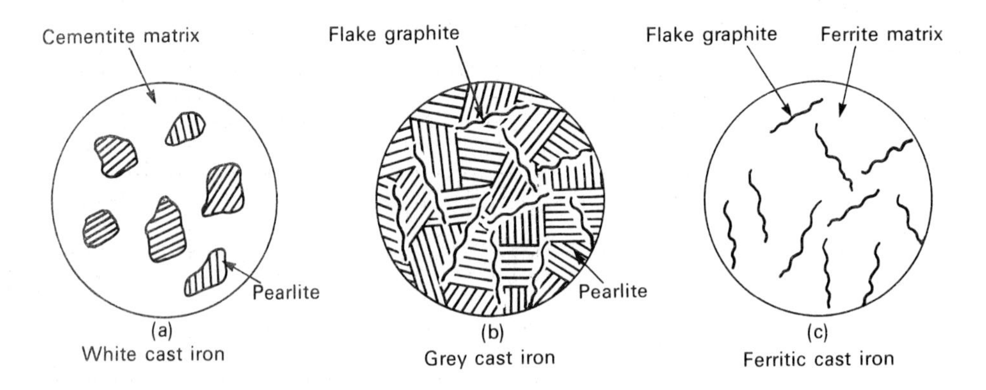

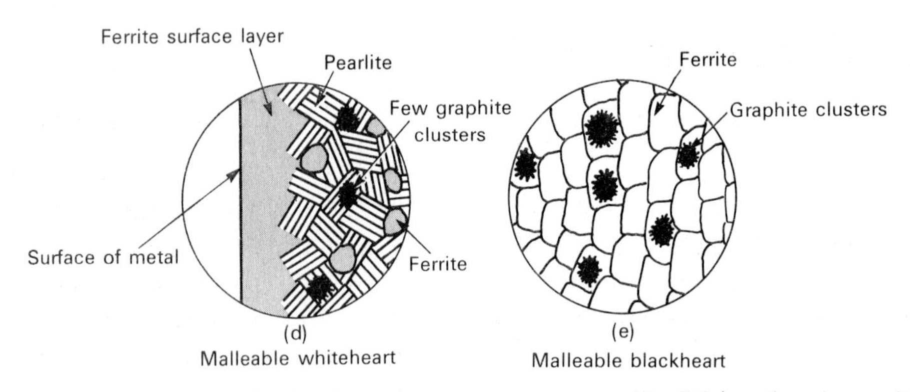

Fig. 5.2 (continued opposite)

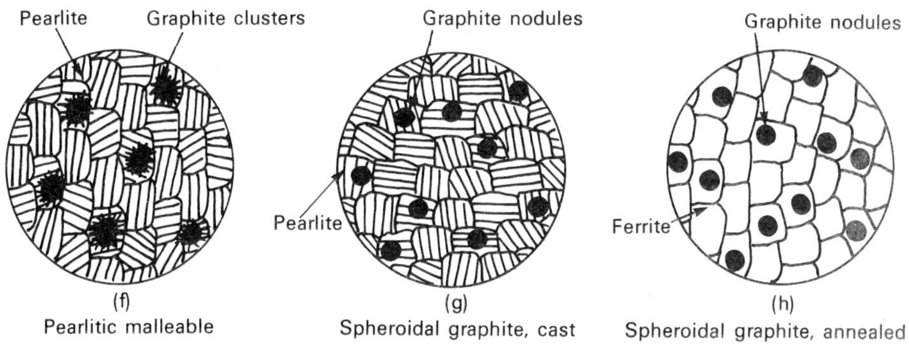

Fig. 5.2 Cast iron structures

irons is usually adjusted to modify the structure so that the graphite occurs as nodules instead of flakes.

Grey iron properties are covered by British Standard BS 1452 (see Table 5.1).

TABLE 5.1 *Specifications and typical properties of cast irons*

Material	Relevant specification	Tensile strength (N/mm^2)	Brinell hardness	Elongation (%)
Grey iron	BS 1452	150–400	100–320	0.2–0.6
Whiteheart malleable	BS 309	340–410	229 max.	3–10
Blackheart malleable	BS 310	290–340	150 max.	6–14
Pearlitic malleable	BS 3333	440–690	150–285	2–7
Nodular	BS 2789	370–800	170–352	2–17

MALLEABLE CAST IRONS

Malleable literally means 'hammerable' – thus malleable cast irons are not brittle.

White cast iron is converted to malleable iron by annealing at a temperature of about 1000 °C in an oxidising atmosphere. The oxidising atmosphere may be obtained either by packing the castings in iron oxide ore or by using a gaseous oxidising atmosphere. This has the effect of decarburising the surface layers of the castings, leaving a ferrite surface layer with a tough pearlite-ferrite core (Fig. 5.2(d)). In very thin sections the decarburisation may extend right through the casting, giving a basically ferrite structure. The resulting products are termed *whiteheart malleable castings* and are produced to British Standard BS 309 with the properties quoted in Table 5.1.

An alternative process consists in heating the white iron castings at about 900 °C in an inert (non-oxidising) atmosphere; for example, by heating them while they are packed in sand. The final structure is ferrite and clusters of graphite (Figs. 5.2(e) and 5.3). The iron is known as *blackheart malleable cast iron*. Typical specification properties from British Standard BS 310 are given in Table 5.1.

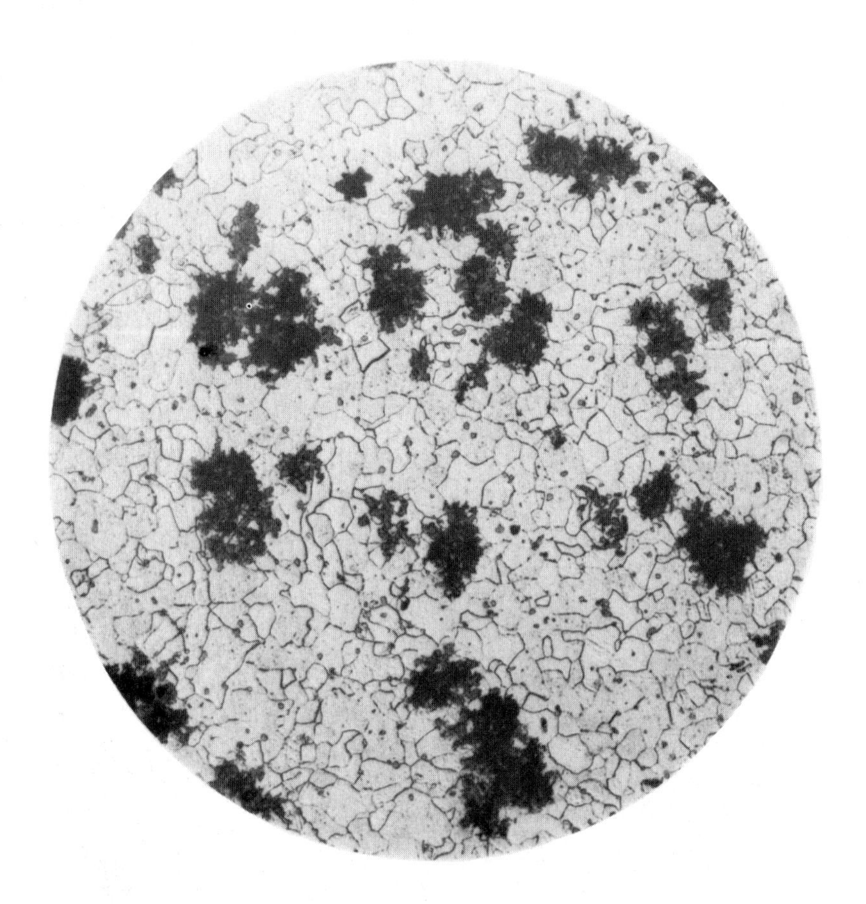

Fig. 5.3 Blackheart malleable cast iron – graphite rosettes in a ferrite matrix (magnification × 400)

Modifications to the 'blackheart' process give a cast iron having a matrix of pearlite instead of ferrite, which has higher mechanical properties (Fig. 5.2(f)). Typical specification properties from British Standard BS 3333 are shown in Table 5.1.

The main advantage of malleable cast irons is that they avoid the presence of flake graphite in the structure and thus produce castings with higher ductility and tensile strength.

NODULAR OR SPHEROIDAL CAST IRONS

The addition of a magnesium alloy to the molten iron just before casting ensures that graphite is present in the form of small spheres (spheroidal) rather than in flake form. The iron is said to be 'inoculated' with the alloy. This 'inoculation' eliminates the stress concentration effect associated with flake graphite and thus gives considerably better mechanical properties. The structure obtained is therefore a matrix of pearlite with nodular or spheroidal graphite (Fig. 5.2(g)). Annealing the alloy at a temperature of 900 °C decomposes the cementite in the pearlite giving a structure consisting of a ferrite matrix with nodular graphite (Figs. 5.2(h) and 4.4). This reduces the tensile strength, but improves ductility. The appropriate British Standard specification is BS 2789, with typical properties shown in Table 5.1.

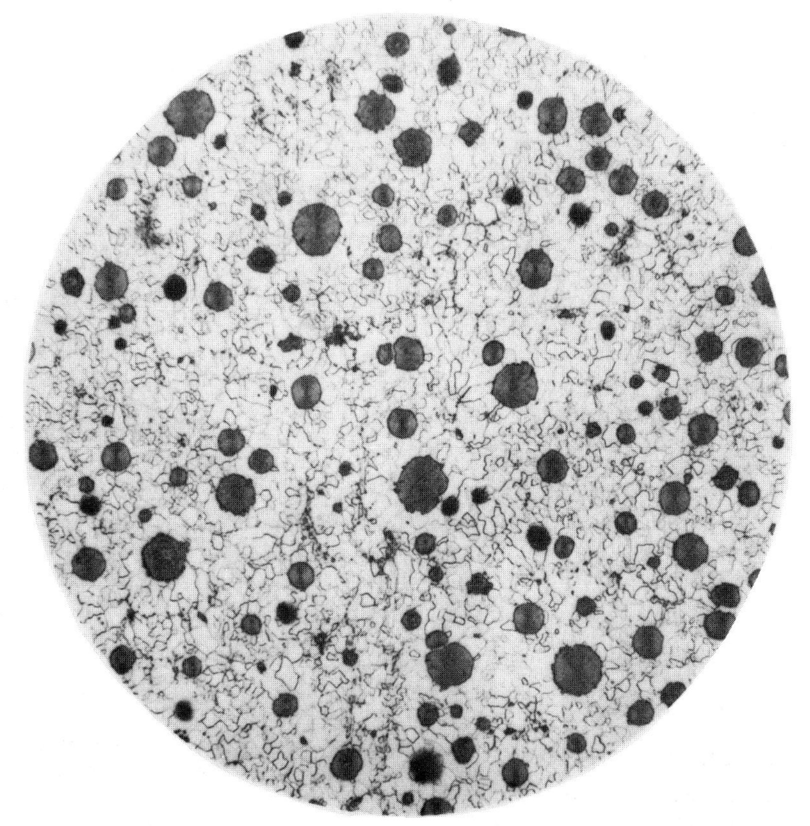

Fig. 5.4 Nodular cast iron – graphite nodules in a ferrite matrix (magnification × 400)

GENERAL APPLICATIONS OF CAST IRON

In general, cast iron is much stronger in compression than in tension. The ultimate compressive stress of grey cast iron, for example, may be as much as four times its ultimate tensile stress. Also the casting process, by its nature, could leave faults in a casting, such as an inclusion of sand, or a void, which would affect its tensile strength much more seriously than its compressive strength.

For these reasons designers try to avoid using cast iron in tension. Where such use is unavoidable, as, for example, when a casting is subjected to bending, the cross-section is usually made unsymmetrical about the neutral axis, as in Fig. 5.5, so that the maximum tensile stress in the material is very much less than the maximum compressive stress.

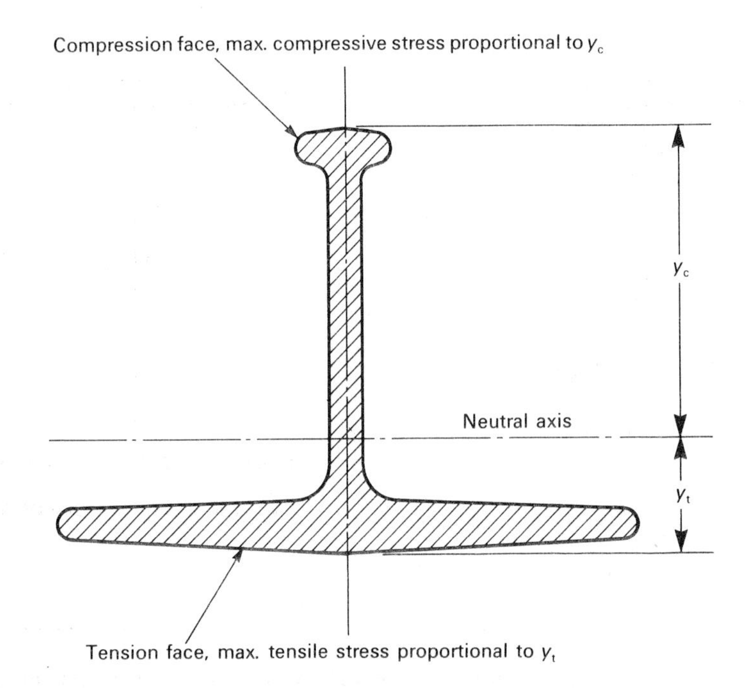

Fig. 5.5 Cross-section suitable for a cast-iron member subjected to bending, i.e. for a material stronger in compression than in tension

The main advantages which cast iron has over steel are:

(i) It is much more fluid than steel when molten, so that intricate shapes such as engine cylinder blocks, which would be difficult or impossible to make in steel, can easily be cast in cast iron.

(ii) The graphite in the cast iron makes it less 'springy' than steel, so that vibrations in cast iron tend to damp out more quickly than they would in steel. Thus it is used in preference to steel for machine tool bed-plates and frames, because it is more effective in damping out cutting tool vibrations ('tool chatter').

(iii) The graphite in the cast iron at a machined surface acts as a solid lubricant. Thus cylinder bores or machine tool slides made of cast iron quickly acquire a polish which makes them much less liable to scuffing or seizure than steel would be.

An outline of the various types of cast iron is shown in Fig. 5.6.

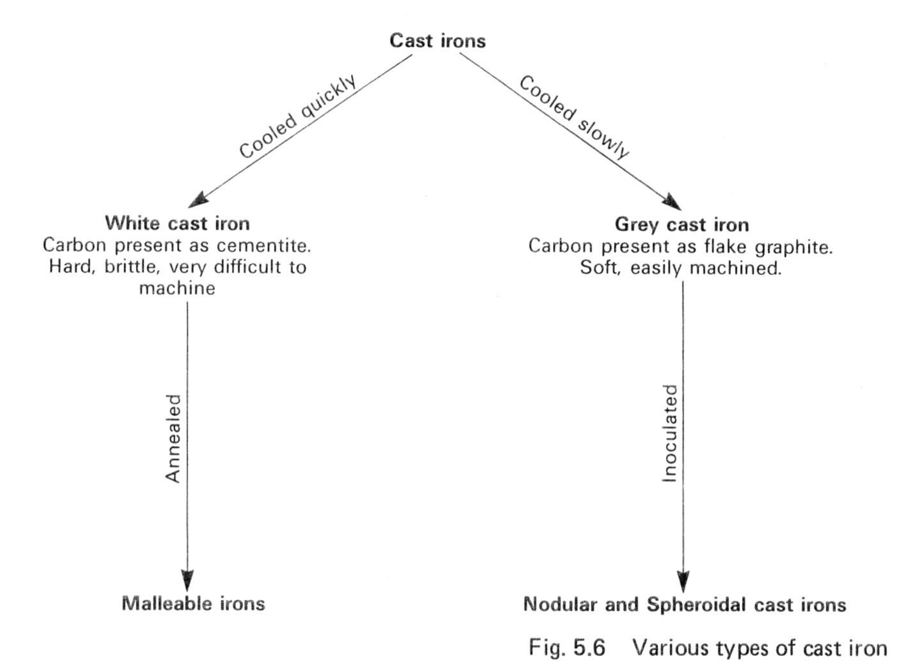

Fig. 5.6 Various types of cast iron

SUMMARY

- Cast iron is iron containing 2–4% C, with traces of other elements. Intricate shapes can be formed by casting, it is a fairly cheap material, and a wide range of properties can be obtained by varying composition, cooling rate and heat treatment.

- The main features of the iron–carbon thermal equilibrium diagram up to 7% C are shown in Fig. 5.7. The eutectic point is at 4.3% carbon and 1140 °C, but the equilibrium diagram is modified slightly if silicon and/or phosphorus are included in the cast iron. To allow for this, an equivalent

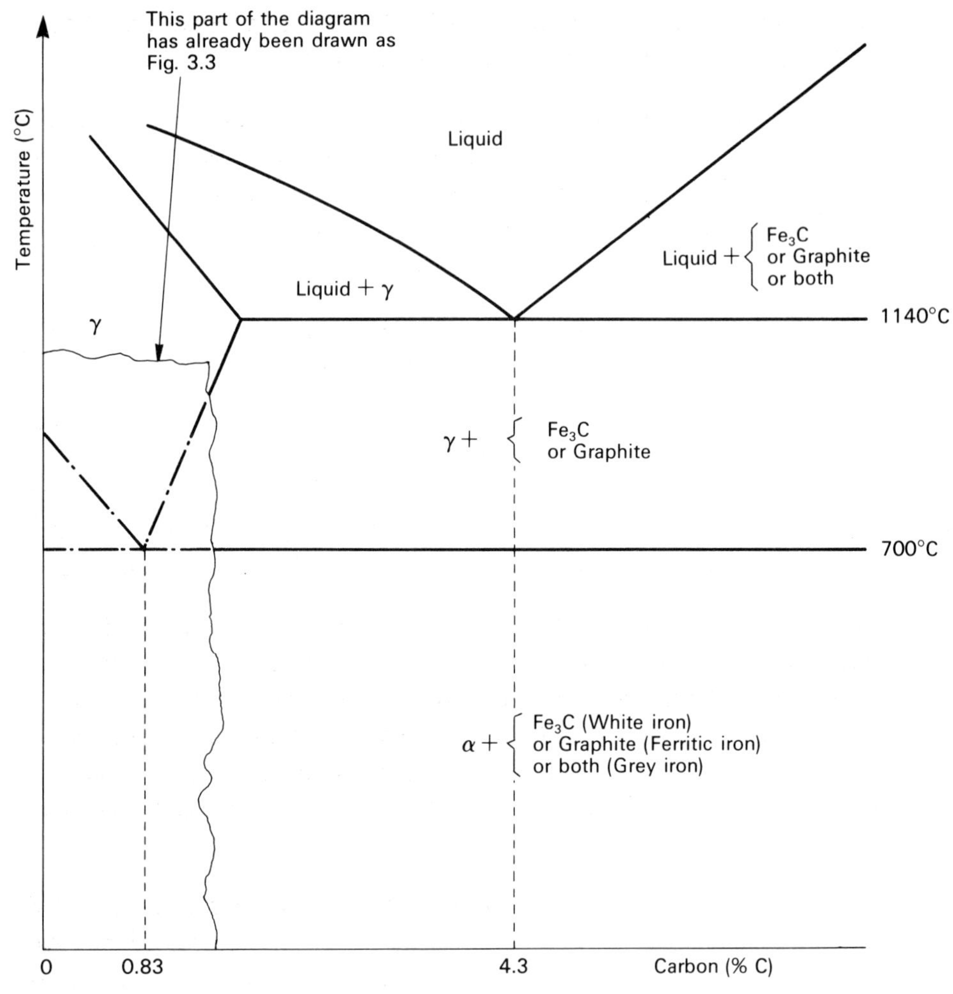

Fig. 5.7 Essential features of the cast iron portion of the iron–carbon equilibrium diagram:
γ = austenite, a solid solution of carbon in face-centred cubic iron – soft, ductile
α = ferrite, a very dilute solid solution of carbon in body-centred cubic iron – soft, ductile
Fe_3C = cementite (iron carbide), a hard brittle iron-carbon compound.
Graphite: a form of pure carbon

percentage of carbon is used instead of the true percentage, when reading from the equilibrium diagram. This *carbon equivalent value* is obtained by adding $\frac{1}{3}$ of the silicon percentage and $\frac{1}{3}$ of the phosphorus percentage to the actual percentage of carbon.

- Only hypo-eutectic (<4.3% C) cast irons are used in industry as hyper-eutectic irons are too brittle.

- *Silicon* which can be present up to 3% breaks down cementite (Fe_3C) into iron and graphite. This gives *grey iron*.

- *Sulphur* has the opposite effect on cementite, stabilising it. As noted already for steel (Chapter 3), sulphur on its own forms brittle iron sulphide in grain boundaries, so it is usually balanced by a similar proportion of *manganese* to form manganese sulphide. Excess manganese improves the tensile strength of the iron.

- *Phosphorus* makes the iron more fluid for casting, but also makes it more brittle, so its percentage is kept low.

- When a cast iron is solidifying, at the eutectic temperature (1140 °C) it forms austenite (1.7% C) and the eutectic mixture of either cementite or graphite. At the eutect*oid* temperature (700 °C) the austenite changes to pearlite. Fast cooling causes cementite to form, rather than graphite, and this gives *white iron*, which is hard, brittle and almost unmachinable.

- Slower cooling gives flake graphite instead of cementite; this iron is grey, soft and easily machinable. Very slow cooling produces a final structure of ferrite and flake graphite called *ferritic iron*. Flake graphite is a source of weakness, causing low tensile strength and brittleness, so before being cast, grey iron may be 'inoculated' with a magnesium alloy so that the graphite appears in rounded instead of flake form. This is called *nodular* or *spheroidal* cast iron.

- In general, white iron can be produced, if required, by the use of 'chills' – cold iron masses in the sand moulds. However, very thick sections have to be mainly grey iron, because the cooling rate of thick sections is slower.

- *Heat treatments*. Nodular cast iron may be annealed at 900 °C to decompose the cementite in the pearlite, giving ferrite + nodular graphite. This reduces the tensile strength but improves ductility.

  White iron may be annealed at 1000 °C in an oxidising atmosphere. This decarburises the outer layer, so that it becomes a layer of ferrite enclosing a tough ferrite–pearlite core which contains some graphite clusters. The resulting iron is called *whiteheart malleable*. Alternatively, white iron may be heated at 900 °C in inert surroundings (e.g. sand) to give *blackheart malleable* iron, consisting of ferrite with clusters of graphite. An improvement to this process leaves pearlite instead of ferrite as the matrix for the graphite clusters. The resulting iron, called *pearlitic malleable* iron, has much better tensile strength.

- *Uses*. Grey iron is used where tensile strength is unimportant, where ease of machining is desirable, and where the flake graphite is useful in providing a solid lubricant for sliding surfaces.

  Malleable irons are used for tough castings, able to withstand impact loads.

  Nodular irons can be produced with even better impact resistances and toughness than malleable irons, and are used where high-quality castings with good machine finish are required.

TABLE 5.2 *Typical applications of cast irons*

Material	Applications
Grey iron	Lathe beds, machine tools, valves, camshafts, pistons, marine propellers, baths, drain pipes, hot water boilers, railings.
Malleable irons	Electrical transmission parts, tools, wheel hubs, pipe fittings, scaffolding fittings, valves, brake shoes.
Nodular cast irons	Crankshafts, gears, earth-moving plant, pipe fittings, impellers, electric motor parts, castors.

For more detailed applications, see the literature of the Council of Ironfoundry Association or of the British Cast Iron Research Association.

  The graphite in cast iron makes it a self-damping material, in which vibrations die out more quickly than in steel.

- *Design of beams*. Because grey iron is weak in tension, cross-sections which carry bending moments are usually made non-symmetrical about the neutral axis, so that the material farthest from the neutral axis, and which therefore has the higher stress, is in compression.

EXERCISE 5

1) State the main advantages of cast irons.

2) Explain the meaning of the term *carbon equivalent* as applied to the composition of cast irons.

3) What are the main elements other than iron and carbon to be found in cast iron?

4) What is the effect of phosphorus and sulphur on the structure and properties of a cast iron?

5) What is the effect of manganese additions to cast irons?

6) Sketch and label the metallographic structures of: (a) a white cast iron; (b) a grey cast iron.

7) How is a white cast iron converted into whiteheart malleable iron? Sketch and label the typical metallographic structure of a whiteheart malleable iron.

8) Why is the presence of flake graphite in the structure of a cast iron undesirable?

9) State a typical application of: (a) a grey iron; (b) a malleable iron; (c) a nodular iron.

10) How may the design of a typical cast iron section allow for the significantly higher compressive strength of cast iron compared with its tensile strength?

6 NON-FERROUS ALLOYS

INTRODUCTION

The term *non-ferrous alloy* refers to any alloy in which the main element is not iron; many of the most important engineering alloys are in this category. There is a very wide range of non-ferrous alloys, but four of the most important systems are those in which aluminium, copper, magnesium or zinc is the main element. As in the case of ferrous alloys (i.e. cast iron or steel), the addition of quite small percentages of alloying elements may considerably modify the properties of the pure metal.

Except for certain specialised applications, it is the alloys rather than the pure metals which are of prime interest to the engineer. This is mainly because the mechanical properties of the alloys are often very much superior to those of the pure metal, for virtually no change in density.

The advantages of a non-ferrous alloy over cast iron or steel may include:

(i) lower density;
(ii) higher electrical conductivity;
(iii) higher thermal conductivity;
(iv) freedom from magnetic effects;
(v) better corrosion resistance under atmospheric conditions;
(vi) better corrosion resistance to specific chemical environments;
(vii) easier fabrication;
(viii) better casting properties;
(ix) better appearance.

The disadvantages may include:

(i) higher cost;
(ii) lower tensile strength;
(iii) higher coefficient of expansion;
(iv) lower melting point;
(v) lower Young's modulus value (so that structures are less rigid);
(vi) need for more costly welding, brazing or soldering techniques.

ALUMINIUM

The main properties of pure aluminium (99.9–99.5% purity) are:

(i) density, 2700 kg/m^3 (approximately one-third that of steel);
(ii) electrical conductivity, approximately two-thirds that of copper;
(iii) thermal conductivity, approximately five times that of steel;
(iv) tensile strength, approximately 50–150 N/mm^2 compared with 300–450 N/mm^2 for mild steel;
(v) Young's modulus, 70×10^3 N/mm^2 (approximately one-third that of steel);
(vi) crystal structure, face-centred cubic;
(vii) corrosion resistance, good – better than carbon steels;
(viii) non-magnetic;
(ix) coefficient of thermal expansion, approximately twice that of steel.

In considering some of these properties it is more realistic to quote the property relative to the density of the material.

For this reason the following values are often quoted:

$$\text{Specific electrical conductivity} = \frac{\text{Electrical conductivity}}{\text{Density}}$$

$$\text{Specific tensile strength} = \frac{\text{Tensile strength}}{\text{Density}}$$

$$\text{Specific Young's modulus} = \frac{\text{Young's modulus}}{\text{Density}}$$

The specific electrical conductivity of pure aluminium is in fact superior to the specific electrical conductivity of pure copper, which has a density of 8900 kg/m^3. Similarly the specific tensile strength and the specific Young's modulus of pure aluminium will be comparable with those of mild steel.

As for all pure metals, the mechanical properties of pure aluminium can not be increased by heat treatment, but the tensile strength and hardness may be increased by cold working (work hardening), for example by drawing or rolling. Cold working results in significant grain distortion, to a degree dependent on the extent of the cold work. Annealing the material after cold work results in recrystallisation, i.e. the reforming of the distorted grains into new small equi-axed grains (Fig. 6.1). The extent of the cold work is often specified thus:

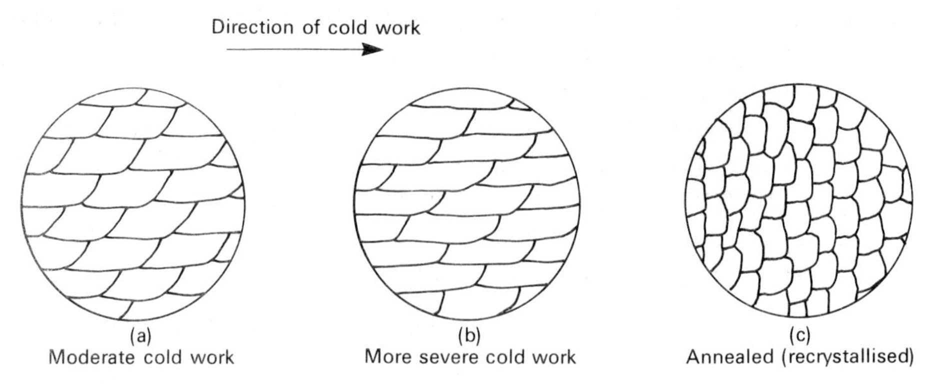

Fig. 6.1 Effect of cold work and annealing on the structure of an alloy

Condition (temper)	Extent of cold work
Annealed (O)	No cold work, recrystallised structure.
$\frac{1}{4}$ hard (H2)	Moderate degree of cold work.
$\frac{1}{2}$ hard (H4)	Significant cold work.
$\frac{3}{4}$ hard (H6)	Severe cold work.
Hard (H8)	Maximum degree of cold work practicable.

Typical applications for commercially pure aluminium products are in situations in which tensile strength is not the significant property. The material is used for food containers, packaging, and chemical processing vessels, where good corrosion resistance and its non-toxic nature are of prime importance. Its high electrical conductivity and low density make it useful for electrical transmission cables.

ALUMINIUM ALLOYS

Aluminium alloys may be divided into two main groups: non-heat-treatable alloys and heat-treatable alloys.

Non-heat-treatable alloys

There are three main alloy systems in this group: the aluminium–$1\frac{1}{4}$% manganese alloys, the aluminium–2-5% magnesium alloys and the aluminium–10-13% silicon alloys. The aluminium–$1\frac{1}{4}$% manganese alloys and the aluminium–2-5% magnesium alloys are primarily available in wrought form (sheet, tubes, forgings, extrusions) and the aluminium–10-13% silicon alloy as sandcastings or diecastings.

The aluminium–$1\frac{1}{4}$% manganese alloy has a higher tensile strength than commercially pure aluminium has, and it resists corrosion equally well. The alloy is easily weldable by gas, arc or resistance methods. It has good ductility and may be readily formed by normal rolling and pressing operations. Typical applications are domestic ware, building panels and some internal aircraft panels where good weldability is an advantage. The mechanical properties of this alloy may be increased by work hardening, and it can be softened by annealing at about 380°C.

The aluminium–2-5% magnesium alloys have higher tensile strengths than the previous group and have exceptionally good resistance to corrosion by sea water. The alloys are subject to work hardening and may be softened by an annealing heat treatment. Prime applications are where exposure to a marine environment is involved; e.g. ship panels, harbour installations.

The most important non-heat-treatable casting alloy is the aluminium–10-13% silicon alloy which has good flow characteristics, enabling pressure-tight castings to be produced. Aluminium and silicon form a eutectic at 11.7% Si (Fig. 6.2). The equilibrium diagram shows that only a small proportion of silicon is soluble in aluminium (the region denoted α on the diagram), and the aluminium–13% silicon alloy would show a structure consisting of plates of primary silicon surrounded by a eutectic mixture of silicon and the dilute solid solution of silicon in aluminium (α). The coarse silicon constituent present as the primary phase and in the eutectic would result in a brittle material. The aluminium–silicon eutectic composition and temperature may be modified either by undercooling (this may be achieved by diecasting thin section castings) or by 'modification' by the addition of about 0.01% sodium to the melt. The effect of this is shown by the dashed lines in Fig. 6.2. After this 'modification' the aluminium–13% silicon alloy will consist of primary (α) solid solution in a fine eutectic mixture, giving a more ductile material.

These two alternative structures are shown in Fig. 6.3. The alloy with the modified structure is used for gearboxes, aircraft and automobile castings, and in chemical processing plant. The castings have good corrosion resistance, although under marine conditions not as good as the aluminium–magnesium alloys.

Heat-treatable alloys

The addition of elements such as copper, magnesium, silicon and nickel often produces alloys which respond to the solution treatment and precipitation hardening types of heat treatment described in Chapter 2. The original Duralumin alloy was based on the aluminium–4% copper alloy, and the aluminium-rich part of the aluminium–copper equilibrium diagram is shown in Fig. 6.4. Considering the 4% Cu alloy on the diagram we see that at 500 °C all the copper will be in solid solution (Fig. 6.5(a)). The slope of the solvus

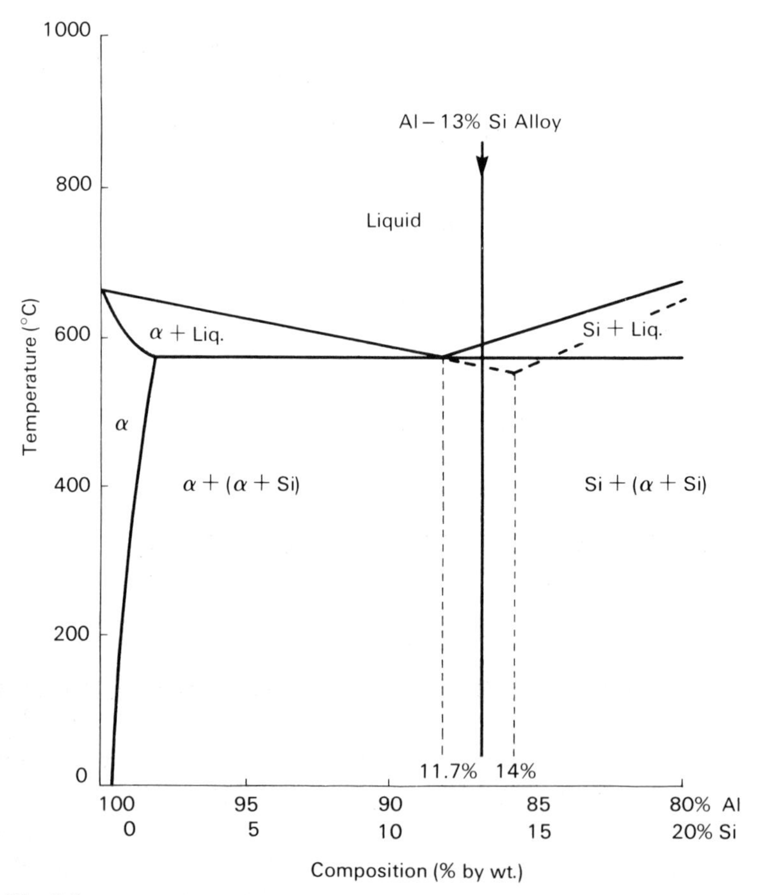

Fig. 6.2 Aluminium–silicon equilibrium diagram. 'Modification' pushes eutectic composition to 14% Si

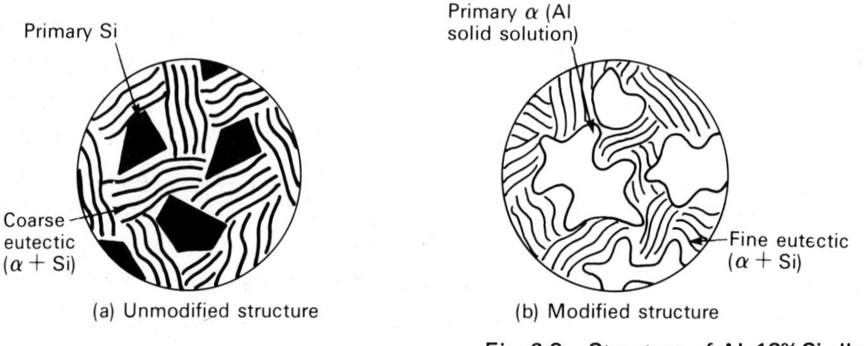

Fig. 6.3 Structure of Al–13% Si alloy

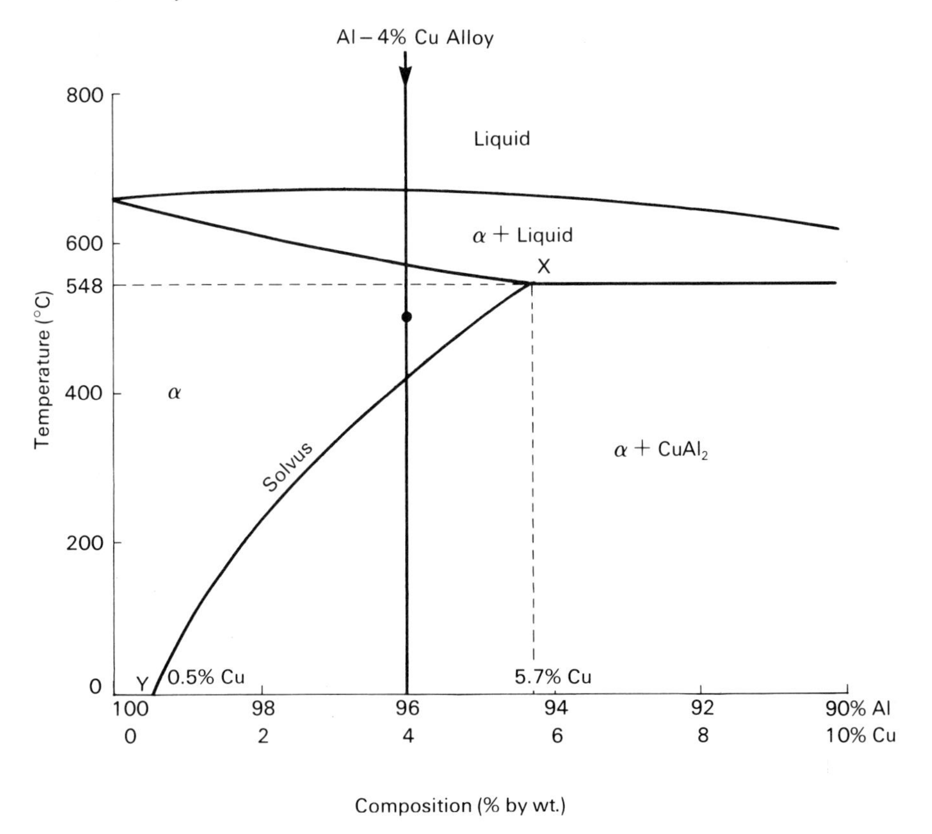

Fig. 6.4 Aluminium-rich end of aluminium–copper equilibrium diagram

XY indicates that the solubility of copper in aluminium decreases as the alloy cools. If the alloy is cooled slowly (air-cooled) from 500 °C, therefore, the grain structure at room temperature will be as shown in Fig. 6.5(b). If, however, the alloy is rapidly cooled from 500 °C (water-quenched), then the precipitation of the $CuAl_2$ compound is retarded, since the precipitation is a diffusion process which is both temperature- and time-dependent. Fig. 6.5(c) shows the grain structure immediately after water quenching. Because it is now a supersaturated solid solution it is unstable, and precipitation of very fine particles of $CuAl_2$ will occur over a period of several days, giving a marked increase in strength but a decrease in the ductility of the alloy. The precipitation of this submicroscopic $CuAl_2$ compound may be accelerated and extended by a low-temperature precipitation-hardening treatment (*age hardening*), by heating for a limited period in the temperature range 120–200 °C. The final structure, revealed under the electron microscope, is as shown in Fig. 6.5(d). The temperature–time combination used for the precipitation treatment is critical since, as Fig. 6.6 shows, it is possible to 'over-age' the material.

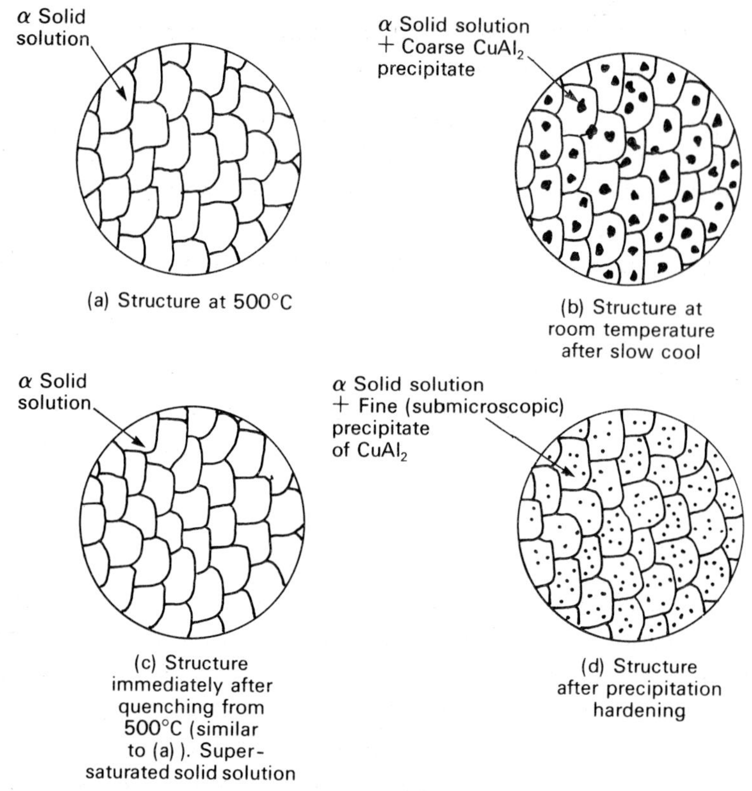

Fig. 6.5 Metallographic structures for Al–4% Cu alloy

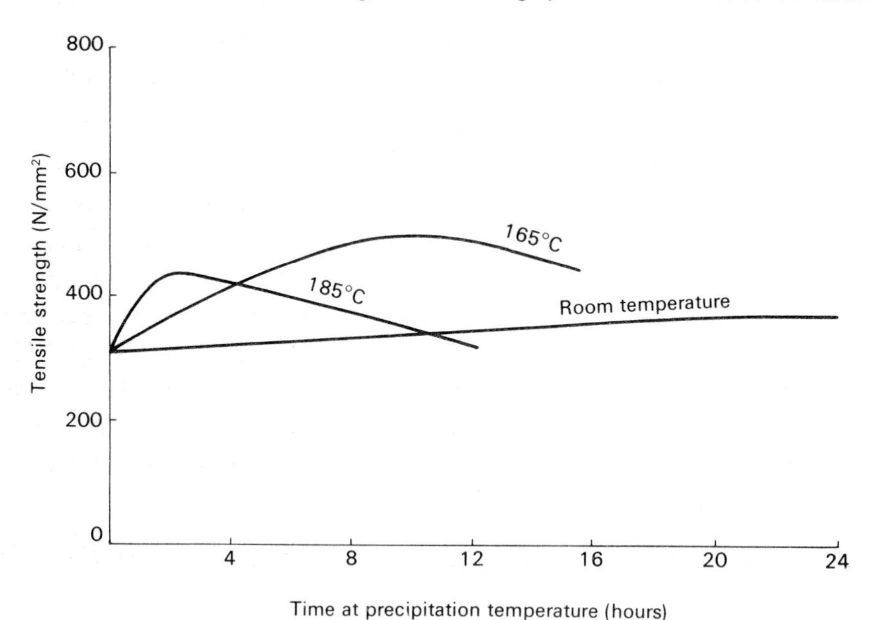

Fig. 6.6 Variation of tensile strength with the temperature and duration of the precipitation-hardening treatments

The advantages of the series of aluminium alloys which respond to 'solution treatment' and 'precipitation hardening' are:

(i) The material may be readily formed (rolled, stretch-pressed, etc.) provided that the forming operation is carried out before significant precipitation occurs – in practical terms, for most alloys, this means within two hours.
(ii) For moderate strength, no further heat treatment is required after the forming operation.
(iii) For maximum strength, only a comparatively low-temperature precipitation-hardening treatment is required after forming.
(iv) For production convenience the material may be stored at sub-zero temperatures immediately after solution treatment, to retard precipitation.

The process schedule is summarised in Fig. 6.7.

A wide range of aluminium alloys which respond to the 'solution treatment' type of heat treatment is available, usually containing magnesium, manganese, silicon, iron or nickel in various amounts. In these complex alloys, various intermetallic compounds are precipitated, e.g. $CuAl_2$, Mg_2Si, $NiAl_3$. The alloy with the highest strength is one containing 5–7% zinc, with the addition of smaller amounts of magnesium, manganese and copper. These very high-strength alloys may, however, be subject to *stress corrosion cracking* (cracking due to a combination of stress and a corrosive environment). The alloys are available in all forms; i.e. as sheet, tubes, extrusions, forgings and castings. A modification to the composition of the aluminium–13% silicon alloy produces a cast alloy which responds to heat treatment.

The heat-treatable aluminium alloys are used primarily where strength is important; e.g. in aircraft structural components, panels, forgings, extruded sections, rivets, pistons and cylinder heads. These alloys have a lower corrosion resistance than either commercially pure aluminium or the work hardening aluminium–$1\frac{1}{4}$% manganese and aluminium–2-5% magnesium alloys. When used in sheet form, therefore, the alloy is often clad with a thin layer of pure aluminium during hot rolling, to give an 'Alclad' alloy. This combines the good corrosion resistance of the outside layer of pure aluminium with the high strength of the core material.

The symbols used to specify the heat treatment condition of an aluminium are:

M condition as manufactured, i.e. no specific heat treatment applied after manufacture

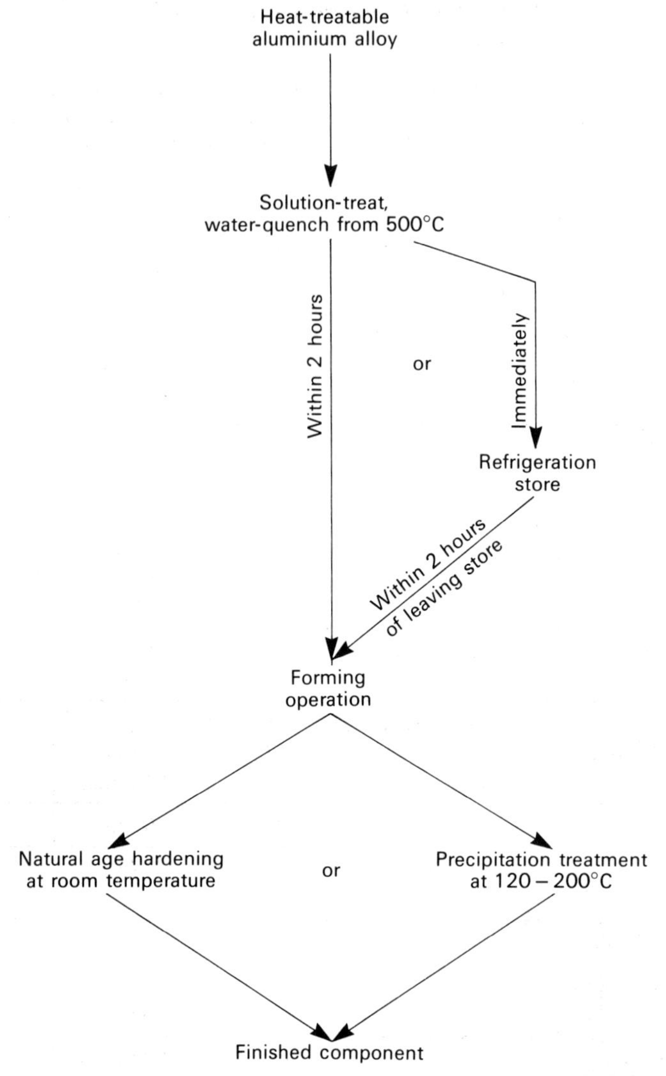

Fig. 6.7 Processing schedule for Al–4% Cu alloy component

O annealed (fully soft) condition

TB solution-treated and naturally aged

TE cooled from an elevated-temperature forming process and precipitation-hardened

TF solution-treated and precipitation-hardened

Typical properties of a range of aluminium alloys are given in Table 6.1.

A summary of the effects of adding magnesium, zinc, copper, manganese or silicon to aluminium is given on p. 122 in the Summary of Chapter 6.

TABLE 6.1 *Properties of aluminium alloys*

Composition	Form	Condition	UTS (N/mm^2)	Elongation (%)
Al 99.99%	Plate/sheet/strip	O H4 H8	65 (max.) 80–95 100 (min.)	30–45 7–12 3–6
Al 99.5%	Plate/sheet/strip	O H4 H8	55–95 100–135 135 (min.)	22–32 4–8 3–4
Al 99.0%	Plate/sheet/strip	O H4 H8	70–105 110–140 140 (min.)	20–30 3–6 2–4
Al + 1% Mn	Plate/sheet/strip	O H4 H8	90–130 140–175 175 (min.)	20–25 3–7 2–4
Al + 3.5% Mg	Plate/sheet/strip	O H2 H4	215–275 245–295 275–325	12–18 5–8 4–6
Al + 4% Cu + 1% Si + 0.8% Mg	Plate/sheet/strip	TB TF	385 (min.) 420 (min.)	13–14 –
Al + 1% Si + 1% Mg + 1% Mn	Plate/sheet/strip	O TB TF	155 (min.) 200 (min.) 295 (min.)	16–18 15 8
Al + 10-13% Si	Castings	M	190 (min.)	5
Al + 9% Si + 4% Cu	Castings	M	180 (min).	1.5
Al + 5% Si + 3% Cu	Castings	M TF	140–160 230–280	2 –
Al 99.5%	Forgings	M	60 (min.)	22
Al + 2% Mg	Forgings	M	170 (min.)	16
Al + 4.5% Mg + 1% Mn	Forgings	M	280 (min.)	12
Al + 4% Cu + 1% Si + 1% Mg	Forgings	TB TF	370–390 435–480	8–13 6–7
Al 99.5%	Bars, extrusions	M	60 (min.)	25
Al + 3.5% Mg	Bars, extrusions	O M	215–275 215 (min.)	16–18 14–16
Al + 4% Cu + 1% Si + 1% Mg	Bars, extrusions	TB TF	370–390 435–480	8–11 6–7
Al + 7% Zn + 2% Mg + 2% Cu	Bars, extrusions	TF	620 (min.)	–

COPPER

The main properties of pure copper are:

(i) density, 8900 kg/m^3 (i.e. heavier than steel);
(ii) electrical conductivity, very good;
(iii) thermal conductivity, very good;
(iv) tensile strength, 150–220 N/mm^2;
(v) Young's modulus, 120×10^3 N/mm^2 (a little more than half that of steel);
(vi) crystal structure, face-centred cubic;
(vii) corrosion resistance, very good;
(viii) non-magnetic;
(ix) coefficient of thermal expansion, approximately 1.5 times that of steel.

Copper is used extensively in the electrical industry because of its high electrical conductivity and excellent corrosion resistance. Electrical applications account for the fact that pure copper is probably used more extensively than any other pure metal. Commercial copper is available in a number of forms, depending on the method of production and subsequent processing, but five of the main grades are given below:

(i) *Tough pitch copper*, which contains up to 0.5% impurities and in particular may contain up to 0.1% oxygen, is produced by the normal fire-refining method. It has a comparatively low electrical conductivity and is difficult to weld. It is cheaper than other forms and may be used for the manufacture of components where high conductivity is not essential.

(ii) *Electrolytic tough pitch copper* is produced from copper which has been obtained by the electrolytic method, i.e. from high-purity (cathode) copper. It has a lower oxygen content (about 0.02%) than tough pitch copper.

Both types of pitch copper are liable to cracking if heated in a reducing atmosphere, for example where hydrogen is present, since the hydrogen diffuses into the copper and combines with the oxygen in the metal, forming water vapour:

$$2H_2 + O_2 = 2H_2O$$

Since the water vapour can not readily diffuse out from the copper, the resulting increase in pressure causes it to crack.

(iii) *Oxygen-free high-conductivity copper* may be produced by melting cathode copper in a non-oxidising atmosphere. It has high conductivity and good forming properties.

(iv) *Phosphorus-deoxidised copper* is produced by using phosphorus to deoxidise molten copper, but any residual phosphorus will reduce the conductivity. It has the advantage over tough pitch copper of being readily welded.

(v) *Arsenical copper* is copper containing about 0.5% arsenic, which gives improved tensile strength and corrosion resistance at high temperatures. The arsenic content does, however, decrease the conductivity of the copper.

Typical applications of the various grades of pure copper are:

(i) tough pitch copper, general fabrication in chemical plant, etc.;
(ii) electrolytic tough pitch copper, important electrical applications such as bus-bars;

TABLE 6.2 *Typical properties and applications of types of copper*

Type	Condition	Tensile strength (N/mm^2)	Elongation (%)	Applications
Fire-refined, or electrolytically refined, tough pitch, high-conductivity copper (99.90% Cu)	Annealed Hard	215 380	55 4	Electrical components.
Oxygen-free high-conductivity copper (99.95% Cu)	Annealed Hard	215 380	60 4	Electrical components. Components requiring brazing.
Tough pitch copper (99.85% Cu)	Annealed Hard	215 380	50 4	General engineering components where high conductivity is not required.
Phosphorus-deoxidised copper (99.85% Cu)	Annealed Hard	215 380	60 4	General engineering where welding or brazing required.
Arsenical copper 99.20% Cu 0.3–0.5% As	Annealed Hard	215 380	50 4	General engineering for use at elevated temperatures. Welding not recommended.

(iii) phosphorus-deoxidised copper, water pipes and boilers;
(iv) oxygen-free high-conductivity copper, electrical applications especially where severe cold forming may be necessary;
(v) arsenical copper, boiler tubes and other high temperature applications; e.g. in a chemical plant.

COPPER ALLOYS

Copper readily forms alloys with a number of elements to give a very wide range of properties. Four of the most important copper alloy systems are:

(i) copper–zinc alloys (brasses);
(ii) copper–tin alloys (bronzes);
(iii) copper–aluminium alloys (aluminium bronzes);
(iv) copper–nickel alloys (cupro-nickels).

Copper–zinc alloys (brasses)

Copper and zinc form a wide range of alloys. The relevant part of the copper–zinc equilibrium diagram is shown in Fig. 6.8. The commercially significant alloys are those containing up to about 50% zinc, since higher zinc contents result in brittle alloys. The significant feature of the rather complex equilibrium diagram is that up to about 35% zinc, the room-temperature structure will consist of a solid solution (α) of zinc in copper, which is very ductile; thus alloys in this composition range may be readily cold-worked. A typical alloy is the alloy known as *cartridge brass*, from its use for making cartridge cases. This is a 70% copper–30% zinc alloy often referred to as 70/30 brass. When this brass has been severely cold-worked, it can suffer from a form of stress corrosion cracking, sometimes referred to as *season cracking*, in which, under corrosive conditions, intercrystalline cracking occurs (Fig. 6.9). The tendency to season-cracking-type failure may be significantly reduced by a low-temperature (300 °C) stress-relief anneal.

Increasing the zinc content to the 35–46% range introduces a second (β) phase into the structure. This β phase, which changes to a modification known as β' at about 450 °C, is comparatively brittle and is a phase based on a copper–zinc intermetallic compound. The change from β to β' at 450 °C is a change from a random to an ordered arrangement of zinc atoms. The introduction of the β' phase reduces the ductility of the brass and therefore limits its ability to accommodate cold working. The $\alpha + \beta'$ alloys are therefore primarily hot-worked at temperatures above 450 °C, when the β' phase is replaced by the more ductile β phase. The most popular $\alpha+\beta'$ is the 60% copper–40% zinc alloy sometimes known as *Muntz metal*.

An important addition of about 3% lead is sometimes made to brasses. The lead, being insoluble, forms lead globules which improve machinability by

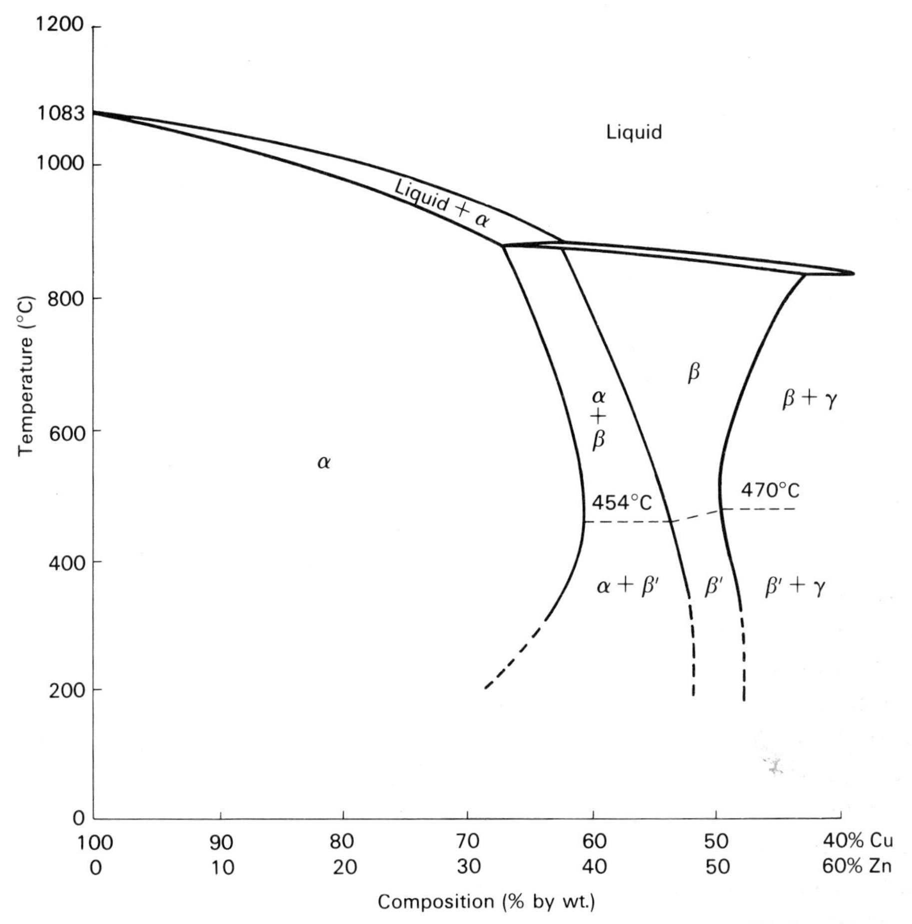

Fig. 6.8 Copper–zinc equilibrium diagram

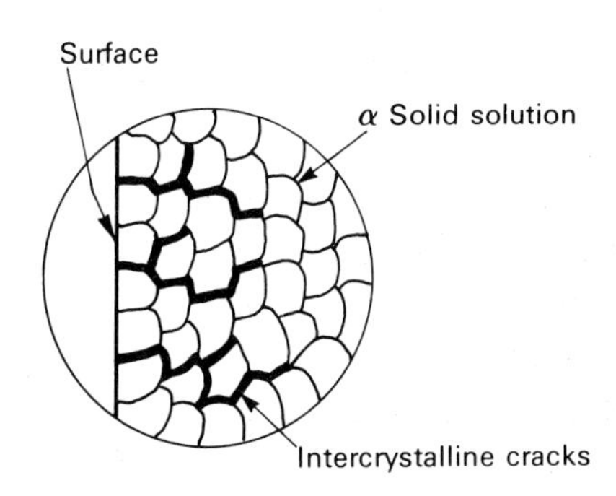

Fig. 6.9 Season cracking of 70/30 brass

aiding the formation of chips during machining operations. Such a brass is often referred to as a *leaded brass.*

Commercial alloys rarely contain more than 46% zinc, since such compositions would produce either single-phase β' or $\beta' + \gamma$ phase structures which are very brittle and have poor corrosion resistance. The tensile strength and ductility as measured by the percentage elongation vary with zinc content as shown in Fig. 6.10. Additions, usually less than 3%, of iron, tin and manganese are sometimes added to brasses to provide specific modifications to properties.

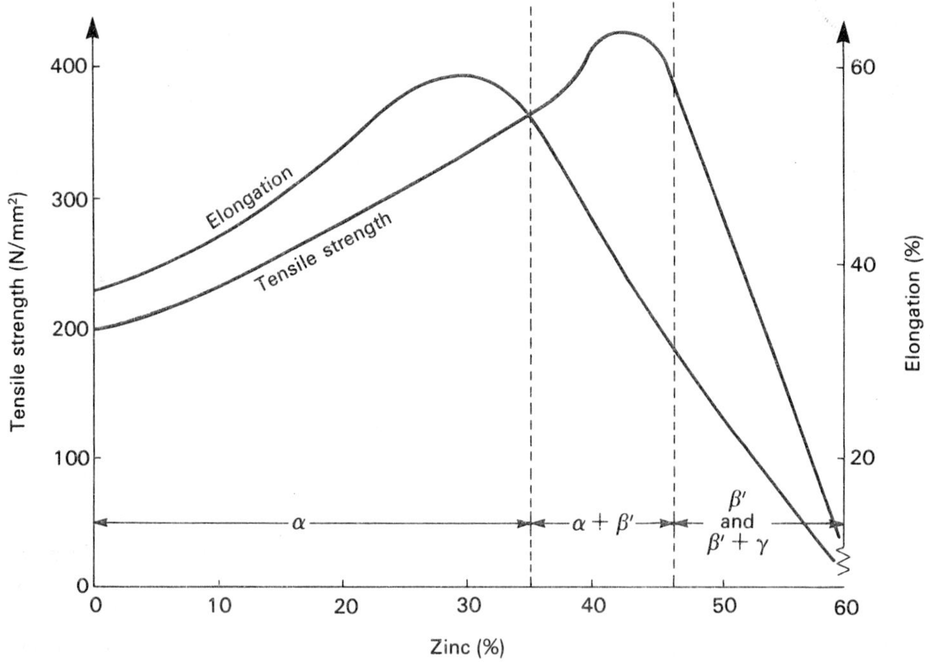

Fig. 6.10 Effect of zinc content on tensile strength and elongation of brasses

Copper–tin alloys (tin bronzes)

The copper-tin equilibrium diagram is shown in Fig. 6.11. Although it looks rather complicated, we are fortunately only concerned with room-temperature structures. The phase boundary indicates a change of solubility in the solid state (see Chapter 2). Under practical cooling conditions, however, the phase boundary from Y is better represented by the dashed vertical line. Thus, under the correct conditions of cooling, alloys containing up to 14% tin may consist of a single-phase (α) structure, whereas alloys containing more than 14% tin will have an $\alpha + \epsilon$ or, in practice, an $\alpha + \delta$ structure.

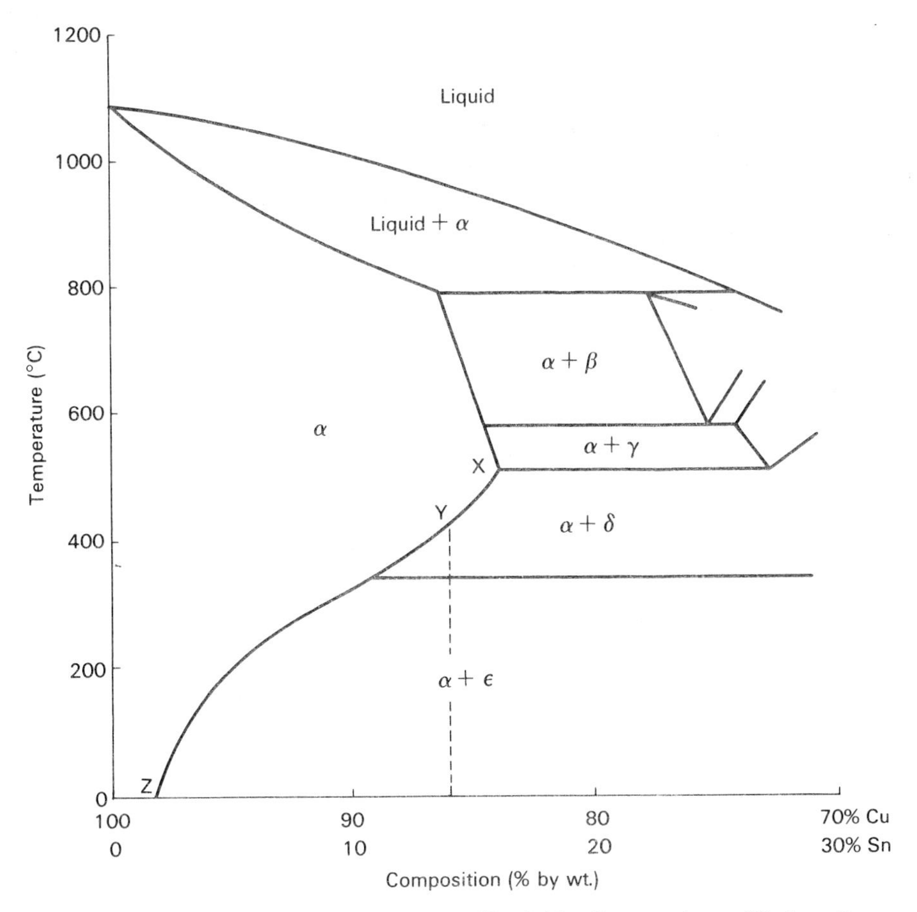

Fig. 6.11 Copper–tin equilibrium diagram

A 95% copper–5% tin alloy will have a single-phase solid-solution (α) structure and will readily cold work. This alloy with 1% zinc added is the standard bronze used for making 'copper' coins, being easily stamped and having good corrosion resistance. Variations may have slightly higher zinc contents and rather less tin.

Alloys with more than about 8% tin have the α + δ structure. The most important of these is *gunmetal*, which has the composition 88% copper–10% tin–2% zinc. It is used in applications requiring good corrosion resistance and moderate strength, such as valve bodies. *Tin bronzes*, which have the composition 85% copper–15% tin, have a structure of soft α and hard α + δ eutectoid; i.e. an ideal bearing alloy structure, consisting of hard particles (α + δ eutectoid) embedded in a soft (α) matrix.

Bronze alloys are often deoxidised with phosphorus during casting and often some phosphorus remains in the structure in the form of the intermetallic compound copper phosphide. This is a hard constituent which improves the wear resistance of the alloy. Such alloys are known as *phosphor-bronzes*. Phosphor-bronze with a composition of 3–9% tin and about 0.2% phosphorus is used in the form of rod, strip and wire to form springs having a good corrosion resistance combined with reasonably high tensile strength.

Copper–aluminium alloys (aluminium bronzes)

The appropriate portion of the copper–aluminium equilibrium diagram is shown in Fig. 6.12. The important feature of this diagram is that a eutectoid reaction occurs at 565 °C at a composition of 11.8% Al. While the α phase is a solid solution of aluminium in copper, the β and γ_2 phases are both based on copper–aluminium intermetallic compounds and will therefore be harder and more brittle than the α phase. Aluminium bronze alloys have very good

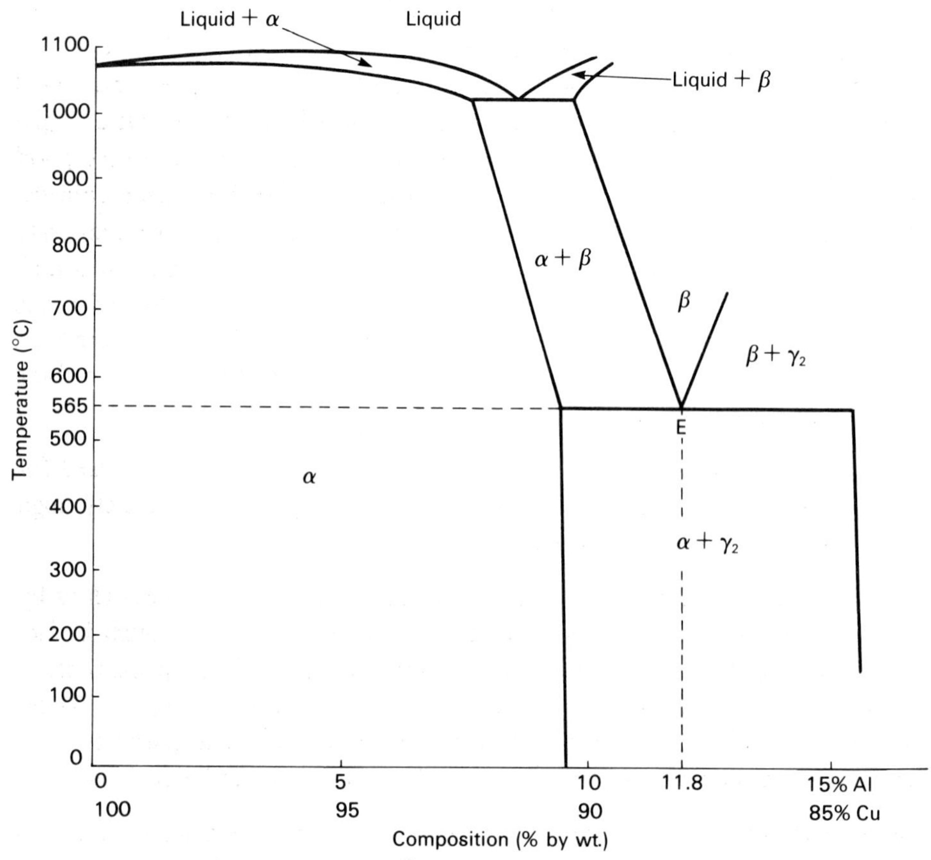

Fig. 6.12 Copper-aluminium equilibrium diagram

corrosion resistance, particularly in marine environments. Typical applications are marine propellers, pumps, valves, condenser tubes and heat exchangers.

There are two main alloy groups:

(i) Those containing less than 9% Al, which will have a single-phase (α) structure and will therefore be suitable for cold working, but will not respond to a hardening-type heat treatment.

(ii) Alloys containing more than 9% Al, which will have the eutectoid $\alpha + \gamma_2$ structure. This will be a coarse, plate-like structure if cooling is slow from above the eutectoid temperature of 565 °C. If, however, the alloy is water-quenched from well above 565 °C, the conversion of the β phase to γ_2 will be suppressed, and a neddle-like martensitic-type structure similar to that obtained in a quenched steel will be obtained. This structure is then modified to give a tougher material by tempering at a temperature of 500–600 °C.

Copper–nickel alloys (cupro-nickels)

The copper–nickel equilibrium diagram is shown in Chapter 2 (Fig. 2.9). We saw that this is a typical solid-solution-type system with copper and nickel being soluble in all proportions. This series of alloys, which all have a single-phase solid-solution structure, are readily cold-worked and have very good corrosion resistance. The 75%Cu–25%Ni alloy is used for UK 'silver' coinage. Because of their excellent corrosion resistance combined with moderate strength and good ductility these alloys are used for marine applications such as valves, pipework and condenser tubes. The 68% nickel alloy with the addition of about 2% iron is known as *Monel metal*. It has good corrosion resistance at moderately high temperatures and is therefore used for turbine blades and valves.

The 60%Cu–40%Ni alloy known as *constantan* or *Eureka* is often used for electrical resistances and for thermocouples to measure moderately high temperatures (up to about 600 °C).

The replacement of a proportion of the nickel in a copper–nickel alloy by zinc provides a series of alloys known as the *nickel silvers*, although in fact the alloys contain no silver, and the name refers to their appearance. A typical composition would be 20% nickel–20% zinc–60% copper. These alloys are used for cutlery which is silver-plated after manufacture. Such items are usually marked by the initials EPNS (electro-plated nickel silver).

Some versions of these alloys have a comparatively high Young's modulus and are used for corrosion-resistant springs.

TABLE 6.3 *Properties and applications of copper alloys*

Type	Condition	Tensile strength (N/mm^2)	Elongation (%)	Applications
70/30 Cartridge brass 70% Cu, 30% Zn	Annealed Hard	320 700	70 5	Deep drawing. Maximum ductility in annealed condition. General cold-forming operations.
60/40 Muntz metal 60% Cu, 40% Zn	Annealed Hard	370 540	45 10	Hot-working alloy. Stampings.
90/10 Gilding metal 90% Cu, 10% Zn	Annealed Hard	280 510	55 4	Imitation jewellery. Architectural work. Readily brazed.
80/20 Gilding metal 80% Cu, 20% Zn	Annealed Hard	310 620	65 5	
Aluminium bronze 95% Cu, 5% Al	Annealed Hard	380 770	70 4	Very good corrosion resistance. Valves, pumps.
Aluminium bronze 90% Cu, 8% Al, 2% Fe	Hot-rolled	380 to 500	40 to 50	Chemical engineering. Suitable for moderate temperatures.
Gunmetal (Admiralty) 88% Cu, 10% Sn, 2% Zn	Sand-cast	250 to 340	12 to 20	Pumps, valves.
Cupro-nickel 80% Cu, 20% Ni	Annealed Hard	340 540	45 5	Excellent cold-working properties. Good corrosion resistance.
Cupro-nickel 70% Cu, 30% Ni, 1% Fe	Annealed Hard	350 650	45 5	Condenser tubes. Excellent corrosion resistance.

MAGNESIUM

The main properties of pure magnesium are:

(i) density, 1740 kg/m^3 – less than a quarter that of steel;
(ii) electrical conductivity, moderate, approximately half that of copper;
(iii) thermal conductivity, three times that of steel;
(iv) tensile strength, moderate;
(v) Young's modulus, 45×10^3 N/mm^2, approximately one-quarter that of steel;

(vi) crystal structure, closed-packed hexagonal;
(vii) corrosion resistance, good;
(viii) non-magnetic;
(ix) coefficient of thermal expansion, three times that of steel.

Probably the most significant feature of magnesium is its very low density. This immediately suggests the possibility of designing very light structures for automobile and aircraft components. The crystal structure, however, is close-packed hexagonal; and, as we saw in Chapter 1, this suggests a material with less ductility than aluminium or copper, both of which have a face-centred cubic unit cell. The comparative low ductility means that magnesium and its alloys may be cold-worked to only a limited extent compared with aluminium.

Magnesium has good corrosion resistance which may be significantly improved by suitable surface treatments.

Powdered magnesium ignites readily, burning with a brilliant white flame, and swarf from the machining of magnesium could be similarly flammable if it is in a fine enough form. However, magnesium in the form of castings, sheet, etc., does not present any fire risk.

Like other pure metals, pure magnesium finds very limited application and it is the alloys of magnesium which are of prime industrial interest.

Magnesium alloys

The most significant property of magnesium alloys is the same as that of pure magnesium: lightness. Their density is 1800 kg/m^3, compared with 2700 kg/m^3 for aluminium and 7900 kg/m^3 for steel. This means that the specific strength (tensile strength/density) of some magnesium alloys is significantly better than that of aluminium alloys or steels. Although magnesium alloys are generally difficult to form at room temperature, they are easily machined and in many instances are readily weldable. Two of the most important alloying elements used in magnesium alloys are aluminium and zinc. The magnesium-rich portion of the magnesium–aluminium equilibrium diagram is shown in Fig. 6.13 and that of the magnesium–zinc equilibrium diagram in Fig. 6.14. From the work covered in Chapter 2 we see that in both cases a eutectic is formed (E in the diagrams) and the phase boundary

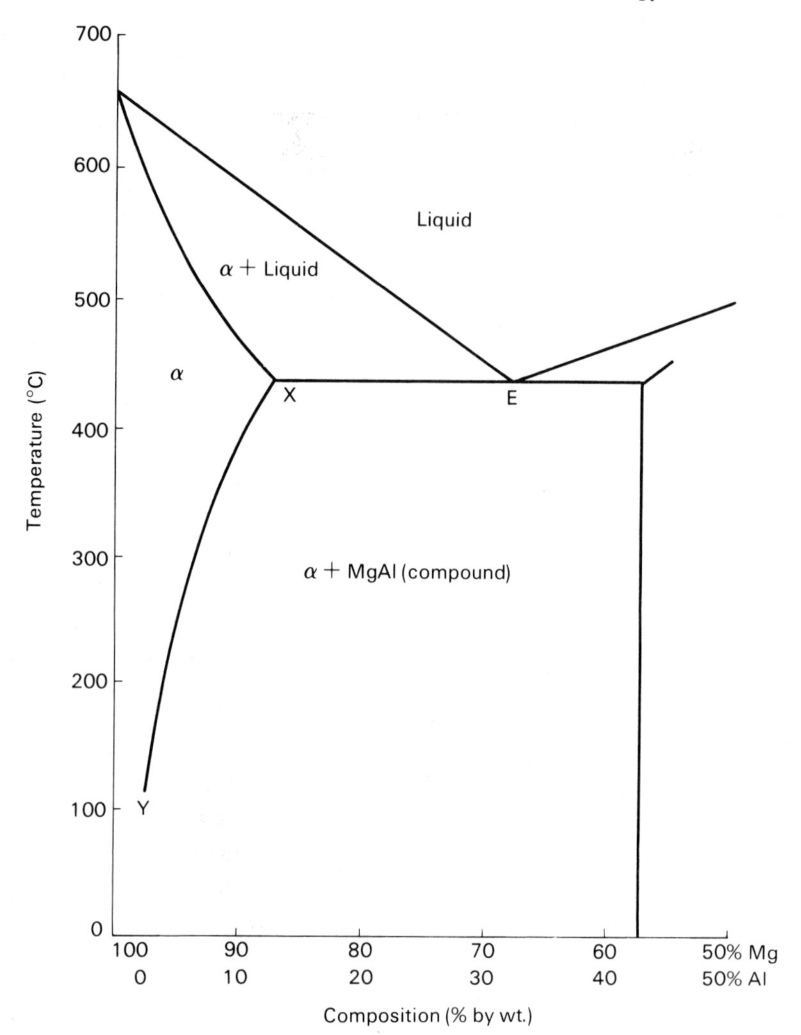

Fig. 6.13 Magnesium-aluminium equilibrium diagram

XY indicates a change of solubility in the solid state. Thus the addition of both aluminium and zinc will produce alloys which respond to the solution-treatment and precipitation-hardening types of heat treatment. Typical treatments are: solution treat from 420°C, and precipitation harden at about 170°C. The alloys are hot-worked in the temperature range 300–350°C, and cooling from the hot-working temperature is sometimes used to replace a separate solution treatment. Such a treatment is sometimes called *press quenching*. Manganese and zirconium are also important alloying elements. Zirconium has a significant grain refining effect on the alloys and gives higher mechanical properties.

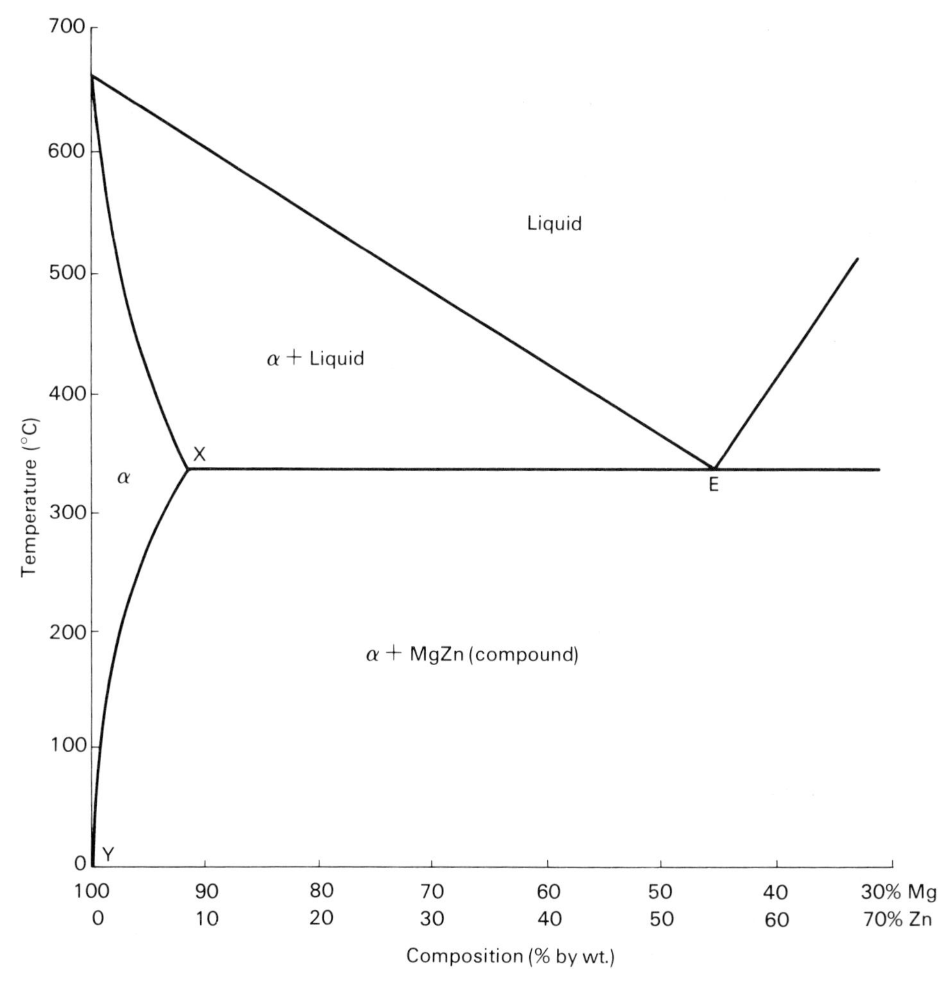

Fig. 6.14 Magnesium-zinc equilibrium diagram

Applications of magnesium alloys include forgings and castings for aircraft structures such as undercarriages and main frame sections. Castings are also used in the automobile industry in such applications as gear cases and wheel hubs. Magnesium alloys are also used in the aircraft industry in the form of sheet and extrusions, for their lightness. Magnesium casting alloys have a very favourable specific strength compared with aluminium casting alloys, since they give similar strength values with lower density. Small castings are used for instrument components and similar applications.

Typical properties of magnesium alloys are given in Table 6.4.

TABLE 6.4 *Typical properties of magnesium alloys*

Casting alloys

Alloy designation and type	Heat treatment condition (see pp. 103–4)	Tensile strength (N/mm^2)		Elongation (%)	
		Sand-cast	Chill-cast	Sand-cast	Chill-cast
MAG 1: Al 8.5%, Zn 0.7% Mn 0.3%	M	140	185	2	4
	TB	200	230	6	10
MAG 4: Zn 3.5–4.5% Zr 1.0%	TE	230	245	5	7
MAG 7: Al 7.5–9.0% Zn 0.3–1.5%, Mn 0.15–0.8%	M	125	170	–	2
	TB	185	215	4	5
	TF	185	215	–	2

Wrought (sheet) alloys

Alloy designation and type	Heat treatment condition	Tensile strength (N/mm^2)	Elongation (%)
MAG-S-101: Mn 1.0–2.0%	M	200	3–5
MAG-S-111: Al 3.0%, Zn 1.0%, Mn 0.3%	M	250	5–7
	O	220–265	10
MAG-S-141: Zn 0.75–1.5%, Zr 0.4–0.8%	M	240	5
MAG-S-151: Zn 2.5–4.0%, Zr 0.4–0.8%	M	250	5–8

ZINC AND ITS ALLOYS

The chief properties of zinc are:

(i) density, 7100 kg/m^3;
(ii) electrical conductivity, less than one-third that of copper;
(iii) thermal conductivity, twice that of steel;
(iv) tensile strength, moderate;
(v) Young's modulus, 90×10^3 N/mm^2, a little less than half that of steel;
(vi) crystal structure, close-packed hexagonal;
(vii) corrosion resistance, good;
(viii) non-magnetic;
(ix) coefficient of thermal expansion, three times that of steel.

Pure zinc, because of its good corrosion resistance, is used extensively as a protective coating on other metals, for example, in galvanising steel. The melting point of zinc is 419 °C compared with 660 °C for aluminium and 1083 °C for copper. Zinc and its alloys make excellent diecasting materials, the alloys most commonly used for this purpose containing up to 5% aluminium and 2% copper. The zinc used to produce the diecasting alloys must be of a very high purity since the presence of even very small quantities of lead, tin or cadmium makes the alloy extremely liable to intercrystalline corrosion, resulting in dimensional changes (growth) and possibly intercrystalline cracking. After casting, the alloys tend to shrink, and if close dimensional control is necessary the castings are 'stabilised' by heating at 100 °C for about 6 hours. The shrinkage is caused by a room temperature precipitation of a constituent of the alloy, and this is taken rapidly to completion by the low-temperature stabilising treatment.

The crystal structure of zinc is close-packed hexagonal and therefore only a very limited amount of cold working is possible (Chapter 1). For this reason zinc sheet is often rolled at a temperature in the range 100–200 °C.

Typical applications of zinc and its alloys are in:

(i) the casings of dry cells used in batteries for torches, etc.;
(ii) automotive components such as door handles and carburettor bodies;
(iii) domestic equipment such as vacuum cleaner components.

Provided such components are being produced in sufficient quantities, the low melting point and excellent casting qualities of zinc alloys enable low unit costs to be achieved.

TABLE 6.5 *Composition and properties of zinc diecasting alloys*

Alloy	Condition	Tensile strength (N/mm²)	Elongation (%)
A {Al: 3.8–4.3%, Mg: 0.03–0.06%}	A A(S)	286 273	15.0 17.0
B {Al: 3.8–4.3%, Mg: 0.03–0.06%, Cu: 0.75–1.25%}	B B(S)	335 312	9.0 10.0

Conditions A and B: alloys unstabilised
Conditions A(S) and B(S): alloys stabilised (100 °C ± 5 °C for six hours)

CHOOSING NON-FERROUS ALLOYS

In selecting a non-ferrous alloy for a particular application a number of factors must be considered, namely:

(i) the design requirements:
   (a) the service environment;
   (b) the service life expected;
   (c) the static strength required;
   (d) the component integrity expected (how critical would a component failure be?);
   (e) any other specific properties required (such as wear resistance, etc.);

(ii) the quantity required (this may vary considerably from a small number of components for an aerospace application, to extensive production runs in the automotive industry);

(iii) the method of fabrication to be used (casting, forging, welding, etc.);

(iv) the availability of material in the required form.

These factors clearly have a significant influence on what is often the primary consideration: the cost of each finished component (the *unit cost*). The importance of the unit cost as a basis of material selection will mainly depend on the nature of the component and its application. In the aerospace industry the component's integrity is often of vital importance, with unit cost a secondary consideration. In the automotive industry, on the other hand, some degree of component integrity may be sacrificed in order to reduce unit costs to levels at which they are competitive with those of other manufacturers.

In Table 6.6 the static tensile strength, density and specific tensile strength (tensile strength/density) of the non-ferrous metals covered in this chapter are compared with mild steel.

TABLE 6.6 *Specific tensile strengths of non-ferrous alloys, compared with mild steel*

Material	Tensile strength (N/mm^2 or MN/m^2)	Density (kg/m^3)	Specific tensile strength (kN m/kg)
Mild steel	400	7900	50
Aluminium	90–600	2700	33–222
Copper	200–1200	8900	22–135
Magnesium	125–350	1740	72–201
Zinc	150–340	7100	21–48

Looking at the values in the right-hand column of the table we can see why aluminium and magnesium alloys are used so extensively in aircraft. The best of those alloys show up very favourably, even when compared with high-tensile steel, which has a specific tensile strength of 165 kN m/kg.

In terms of corrosion resistance, the order of merit of the alloys will depend on the nature of the service environment. Of the non-ferrous metals, under atmospheric conditions, unprotected aluminium and magnesium have the least resistance to corrosion. The corrosion resistance of both of these alloys may, however, be significantly improved by surface treatments, so that for most applications, corrosion resistance under atmospheric conditions is not a limiting factor. Some copper, aluminium and magnesium alloys are apt to suffer from stress corrosion cracking, however. As explained earlier, this is a tendency for intercrystalline cracks to develop due to the combination of stress and a corrosive atmosphere.

Ease of fabrication is a further important consideration. The following factors may affect the choice of material from this point of view:

(i) Some copper and zinc alloys have to be worked in the heated state – this increases production costs.
(ii) Machinability varies considerably – in general a grain structure which aids the formation of small 'chips' is best. This is sometimes arranged by adding an element such as lead to the material, or by suitable heat treatment.
(iii) When forming has to be done by means of bending or stretching operations, materials in the annealed condition will be most easily formed.
(iv) Some of the higher-strength aluminium and magnesium alloys are difficult to weld by conventional methods.
(v) Stress-relieving heat treatments are often required after welding.
(vi) Zinc alloys are difficult to solder.

Turning, finally, to choice of material from the point of view of raw material costs, Table 6.7 shows the cost per unit mass and cost per unit volume of the main non-ferrous materials, relative to mild steel. However, the cost of the

TABLE 6.7 *Approximate ratios of costs for raw material in the form of ingots*

Material	Cost per unit mass relative to mild steel	Cost per unit volume relative to mild steel
Mild Steel	1	1
Aluminium	5.4	1.85
Copper	8.7	9.8
Magnesium	12.2	2.7
Zinc	3.1	2.8

raw material may well be an insignificant part of the unit cost of the finished component. The unit cost of the component is made up of material costs, capital and other overhead costs, and labour costs. These factors will vary considerably for different components and for different industries.

SUMMARY

- The properties of commercially pure aluminium, zinc, magnesium and copper are compared in Table 6.8.
- The properties of *aluminium* may be greatly modified by the addition of small percentages of magnesium, zinc, copper, manganese or silicon, as described below.

  Aluminium + magnesium. Less than 5% Mg provides a higher-strength work-hardening alloy with particularly good corrosion resistance. Above 6% Mg gives a series of heat-treatable alloys. In combination with silicon, Mg contributes to the precipitation-hardening mechanism of some alloys.

  Aluminium + zinc. Proportions of 5% to 7% Zn give the highest strength of any aluminium alloy. This type of alloy could suffer from stress corrosion cracking, however.

  Aluminium + copper. 4% Cu gives an alloy most responsive to solution treatment and precipitation hardening. This is the basis of Duralumin (Dural), the most widely used material for aircraft. Cu reduces the corrosion resistance and ductility of aluminium, but gives a big increase in proof stress and ultimate tensile stress.

  Aluminium + manganese. 2% Mn gives an alloy with almost as good forming properties as commercially pure aluminium, with good corrosion resistance and a higher tensile strength.

  Aluminium + silicon. Si provides good casting properties. It gives a wide range of properties with good corrosion resistance.

  Solution treatment and precipitation hardening applies to various alloys formed when small percentages of copper, magnesium, silicon or nickel are added to aluminium, but especially relates to the Al + 4% Cu type of alloy known as *Duralumin*. At 500 °C the copper is in solid solution in the aluminium. The solubility of the copper in aluminium becomes less than 4% as the solution cools, so a coarse precipitate of $CuAl_2$ appears if the alloy is allowed to cool naturally, and the material is then in the annealed state – soft and ductile. If the material is water-quenched from 500°C, however, the copper is retained as a supersaturated solid solution. This is called *solution treatment*. In this state the material

TABLE 6.8 *Comparison of the properties of the main non-ferrous elements*

Element	Density (weight)	Conductivities		Tensile strength[1]	Deflection under load[2]	Crystal structure[3]	Corrosion resistance	Magnetism	Coefficient of thermal expansion
		Electrical	Thermal						
Aluminium	$\frac{1}{3}\times$ steel	$\frac{2}{3}\times$ copper[4]	5 × steel	$<\frac{1}{2}\times$ mild steel	3 × steel	f.c.c.	Good	Non-magnetic	2 × steel
Zinc	Nearly as heavy as steel	$<\frac{1}{3}\times$ copper	2 × steel	Moderate	$>$ 2 × steel	c.p.h.	Good	Non-magnetic	3 × steel
Magnesium	$\frac{1}{4}\times$ steel	$\frac{1}{2}\times$ copper	3 × steel	Com-parable with aluminium	4 × steel	c.p.h.	Good	Non-magnetic	3 × steel
Copper	Heavier than steel	Very good	Very good	About half × mild steel	Nearly 2 × steel	f.c.c.	Very good	Non-magnetic	1.5 × steel

Notes: (1) Tensile strengths are very approximate, as they depend on the amount of cold working.
(2) Comparisons of deflections are the inverse of Young's modulus comparisons – see introduction.
(3) f.c.c. means *face-centred cubic* – hence ductile material.
c.p.h. means *close-packed hexagonal* – hence brittle material.
(4) Entries such as '$\frac{2}{3}\times$ copper' or '4 × steel' should be read as 'two-thirds of that of copper' and 'four times that of steel'.

behaves as annealed, and can be easily formed. After about two hours, a much finer precipitate of $CuAl_2$ starts to appear, the process continuing for several days. When it is complete, the material is a good deal stronger but less ductile. The process can be accelerated and extended by reheating to about 165 °C for about ten hours (time and temperature are fairly critical) immediately after the forming process; this gives even greater strength than natural ageing does. Sub-zero temperatures halt the natural age-hardening process in this type of material, so the material may be put into a refrigerator immediately after quenching, and stored until it is required for forming.

- Pure *zinc* is used to give a corrosion-resistant coating to steel (galvanising) though such a coating has only a limited life.

  A useful application of zinc is in zinc-based diecasting alloys. A typical alloy contains up to 5% Al, up to 2% Cu, with traces of other metals as impurities. Lead, tin and cadmium must be eliminated from the alloy as far as possible, as they cause corrosion at crystal boundaries, leading to distortion and surface cracks.

  The low melting point of zinc (419 °C) allows the alloy to be forced under pressure into smooth steel moulds (dies) to give an instant casting with a 'machined' finish, so that further machining is usually unnecessary. Thus although the material is about three times as expensive as steel, the finished parts are cheaper, provided production quantities are sufficient to carry the cost of the dies. The process is similar to the moulding of polymer materials, but gives a much more rigid and hard-wearing product.

  A possible disadvantage of zinc-based diecastings is a tendency to shrink after being cast, due to the gradual precipitation of a constituent of the alloy. This process can be brought to rapid completion by *stabilising annealing*: heating at 100 °C for about six hours.

  Typical applications of diecasting are carburettor parts, printing devices (with characters formed by casting) and gears (with teeth formed by casting). Diecastings can also include ready-made 'drilled' holes and internal and external screw-threads.

- *Magnesium* is usually alloyed with small percentages of zinc or aluminium, and because of the reduced solubility of these elements in magnesium at room temperature, some of the resulting alloys can be solution-treated and precipitation-hardened like Duralumin. Magnesium alloys may also include manganese or zirconium, the latter for grain refining and higher mechanical properties.

  Magnesium alloys can be cast or forged; they machine easily and in some cases can be welded. Because of their low melting point, they can be diecast. They are also available as sheet or extrusions. They are chiefly used for aircraft and automobile components, because of their lightness.

- A great deal of commercially pure *copper* is used for conductors in electrical machines and electricity distribution networks, and for pipe-work where corrosion resistance and ductility are desirable. Where the copper is to be used in a hot atmosphere containing combustible gases it should be free from oxygen, or products of combustion will form within the copper, causing cracks.

  Tough pitch copper contains up to 0.5% impurities, including up to 0.1% oxygen. It is relatively cheap but has lower conductivity and is difficult to weld.

  Electrolytic tough pitch copper is much purer, so is a better conductor, and has much less oxygen (0.02%).

  Oxygen-free high-conductivity copper is electrolytic (therefore high-purity) copper in which all subsequent melting has been in a non-oxidising atmosphere. It has high conductivity and good ductility when annealed.

  Phosphorus-deoxidised copper: the phosphorus remaining in the metal reduces conductivity but aids welding.

  Arsenical copper contains about 0.5% As, which improves tensile strength and corrosion resistance but decreases conductivity.

- *Copper–zinc alloys* (*brasses*):
  (i) *Cartridge brass* (70% Cu–30% Zn), very ductile but may suffer from stress corrosion after severe cold working – this can be reduced by stress relief anneal at 300 °C.
  (ii) *Muntz metal* (60% Cu, 40% Zn) has less ductility and is therefore usually hot-worked (stampings).

  3% lead may be added to brasses to improve machinability.

- *Copper–tin alloys* (*bronzes*):
  (i) *Gunmetal* (88% Cu–10% Sn–2% Zn) gives castings with good corrosion resistance and moderate strength.
  (ii) *Tin bronze* (85% Cu, 15% Sn) is used for bearing bushes, having hard particles ($\alpha + \delta$ eutectoid) in softer (α) material.
  (iii) *Phosphor-bronze* containing 3–9% Sn and 0.2% P is deoxidised with phosphorus during casting, the remaining phosphorus forming hard copper phosphide particles in softer bronze material. It is therefore used for bearing bushes, and having good corrosion resistance and tensile strength it is also used for springs.

- *Copper–aluminium alloys* (*aluminium bronzes*) are suitable for shipbuilding castings (propellers, pumps, etc.):
  - (i) *< 9% Al.* Very good corrosion resistance. Can be cold-worked but cannot be hardened by heat treatment.
  - (ii) *> 9% Al.* Can be quenched from well above eutectoid temperature (565 °C) and tempered at 500–600 °C.

- *Copper–nickel alloys* (*cupro-nickels*). Copper and nickel are completely soluble in each other in all proportions, in the solid state, and thus produce single-phase alloys which are readily cold-worked. These have very good corrosion resistance, reasonable strength, and high melting point, so are used for coinage and marine and high-temperature applications.
  - (i) *Monel metal* (30% Cu, 68% Ni, 2% Fe) is used for turbine blades and valves.
  - (ii) *Constantan or Eureka* (60% Cu, 40% Ni) is used for electrical resistances and thermocouples.
  - (iii) *Nickel silver* (60% Cu–20% Ni–20% Zn) is a silvery alloy used for EPNS cutlery.

- *Choosing non-ferrous alloys.* Factors to be considered include:
  - (i) constraints imposed by environment and usage, such as corrosion resistance, wear resistance, duration of useful service, and reliability;
  - (ii) quantity to be produced;
  - (iii) method of production;
  - (iv) availability of material in required form;
  - (v) unit cost (including material costs, labour costs, and a share of the capital cost of any special production equipment required). The cheapest materials do not necessarily give the lowest unit costs: a more expensive alternative material may permit manufacture by a less labour-intensive method (e.g. diecasting).
  - (vi) When lightness is important, specific tensile strengths (tensile strength/density) should be compared – see Table 6.6 on p. 120.

  Material costs per unit mass and per unit volume are compared in Table 6.7 on p. 121.

EXERCISE 6

1) State five possible advantages of non-ferrous alloys over steels.

2) State five possible disadvantages of non-ferrous alloys compared with steels.

3) What would you infer regarding the ductility and forming characteristics of aluminium compared with steel from the fact that the crystal structure of aluminium is face-centred cubic?

4) Explain the meaning of the terms *specific tensile strength* and *specific Young's modulus.*

5) State the main properties of the aluminium–$1\frac{1}{4}$% manganese alloy.

6) What are the main properties and applications of the aluminium–5% magnesium type alloys?

7) Describe how the 'modification' process changes the metallographic structure of the aluminium–13% silicon alloy.

8) State the advantages of the aluminium alloys which respond to solution treatment.

9) With the aid of a diagram explain the process schedule for a heat-treatable aluminium alloy sheet metal component subject to a fairly severe press forming operation.

10) With the aid of a diagram show the effect of over-ageing on the tensile strength of a heat-treatable aluminium alloy.

11) What is the difference between *tough pitch* copper and *electrolytic tough pitch* copper?

12) Why should the use of either type of copper referred to in Question 11 be avoided in a hydrogen-containing atmosphere?

13) What is the effect of small additions of arsenic to copper?

14) What is the effect of increasing zinc content on the strength and ductility of copper–zinc alloys?

15) What is meant by the term *season cracking* applied to 70/30 copper–zinc alloys? How may the tendency to *season cracking* be reduced?

16) How may the machinability of a brass be improved?

17) State the composition and properties of gunmetal.

18) The 75% copper–25% nickel alloy is used for UK 'silver' coinage. What properties make it suitable for this application?

19) Alloys of copper and nickel always exhibit a single-phase structure. Explain this statement.

20) What are the properties of magnesium alloys which make them attractive to the aerospace engineer?

21) Why is the purity of zinc diecasting alloys very critical?

22) Why is a *stabilising* heat treatment carried out on zinc alloy castings?

23) State the main factors to be considered in selecting a non-ferrous alloy for a specific application.

24) What factors must be considered in arriving at a unit cost figure for a non-ferrous component? Why may this figure be of less significance to the aerospace engineer than to the automotive engineer?

25) State the factors to be considered in assessing the *ease of fabrication* of a non-ferrous component.

7 POLYMERS

POLYMER STRUCTURES

Carbon-based molecules

Polymers are materials whose molecules are built up from a series of smaller units. The most common ones are based on the carbon atom.

A carbon atom has an outer 'shell' which contains four electrons and has vacancies for four more (see Chapter 1 and Appendix 2). It readily bonds chemically to four atoms with a vacancy for one electron each in their outer shells. This type of bonding (*covalent bonding*) is due to the sharing of one electron between two atoms, so that at any instant either atom could be considered to have a full complement of electrons in its outer shell. Elements whose atoms are suitable for bonding to carbon in this way include hydrogen, fluorine and chlorine.

A carbon atom with two outer shell vacancies filled by this kind of bonding can also bond to a similar carbon atom on either side. This enables the formation of the long chain molecules we call *polymers*.

An example of a simple carbon-based molecule is *ethane*, a compound of carbon and hydrogen shown diagrammatically in Fig. 7.1. In this case all four covalent bonds to each carbon atom are satisfied.

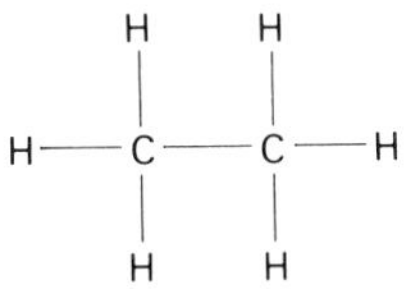

```
    H   H
    |   |
H — C — C — H
    |   |
    H   H
```

Fig. 7.1 Diagram of the ethane molecule. Each letter represents one atom of the corresponding element. The dashes indicate covalent bonds

In some types of molecule, all four covalent bonds may not be satisfied, and adjacent carbon atoms then form a double bond to each other, as in the ethylene molecule, shown in Fig. 7.2.

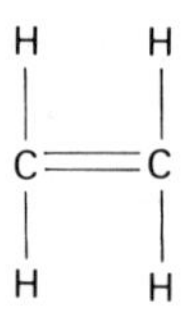

Fig. 7.2 Ethylene – only three of the carbon covalent bonds are satisfied, so a double bond forms between the carbon atoms

The double bond is more readily broken than a single bond, and this means that the ethylene molecule will be more reactive than (say) the ethane molecule.

A substance such as ethylene, consisting of single molecules, is called a *monomer*. It may be modified by breaking the double bonds to allow the simple monomer molecules to join up to form extended molecular chains characteristic of a polymer material. The formation of polyethylene (polythene) from ethylene in this way is illustrated in Fig. 7.3. This process is known as *polymerisation*.

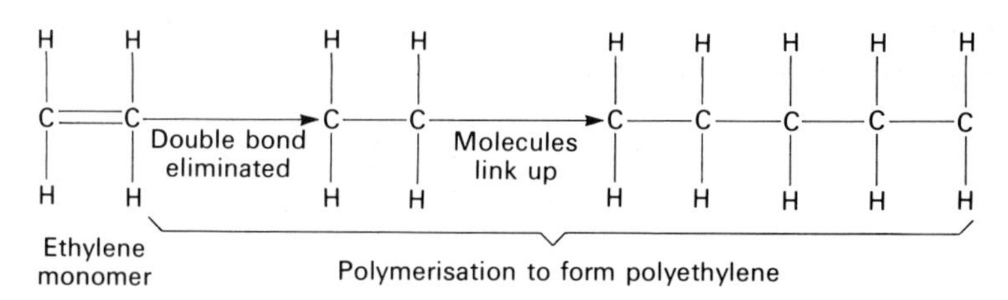

Fig. 7.3 Formation of polythene

This type of polymerisation process produces a long chain-like molecular structure, which in practice may be helical in form, rather like a spring. The long chain molecules with their helical geometry tend to give such polymers marked flexibility. Another example of the formation of a chain-like linear polymer is the formation of the polytetrafluoroethylene* (PTFE) molecule from the simple tetrafluoroethylene monomer as in Fig. 7.4.

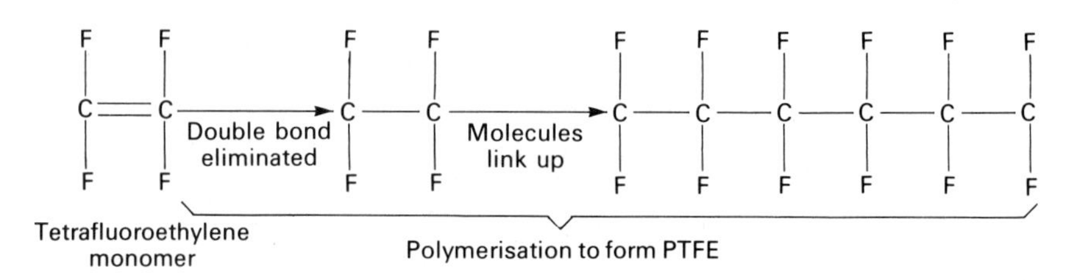

Fig. 7.4 Formation of polytetrafluoroethylene

*The name is a combination of the words *poly* (many), *tetra* (four), *fluoro* (fluorine) and *ethylene*.

Ethane, in which all four carbon bonds are satisfied, is known as a *saturated hydrocarbon*. Ethylene, on the other hand, is an example of an *un*saturated hydrocarbon, in which a double bond is necessary. A further important example of an unsaturated monomer is vinyl chloride which may be polymerised to form the linear chain-like polyvinylchloride (PVC) molecule, as shown in Fig. 7.5.

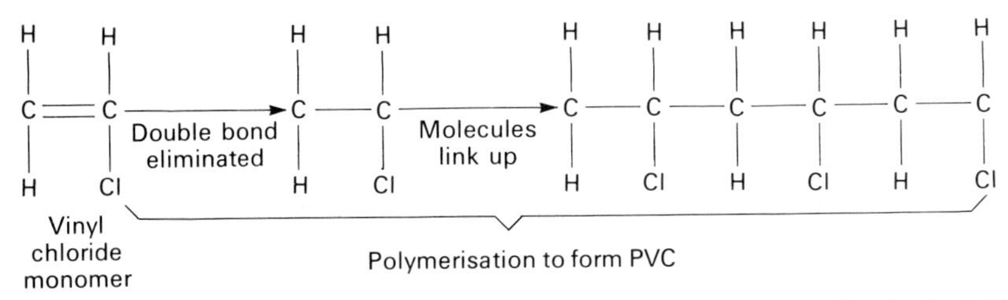

Fig. 7.5 Formation of polyvinylchloride

Linear, branched and cross-linked molecular chains

In the previous section we have met linear molecular structures, which are of the general form shown in Fig. 7.6.

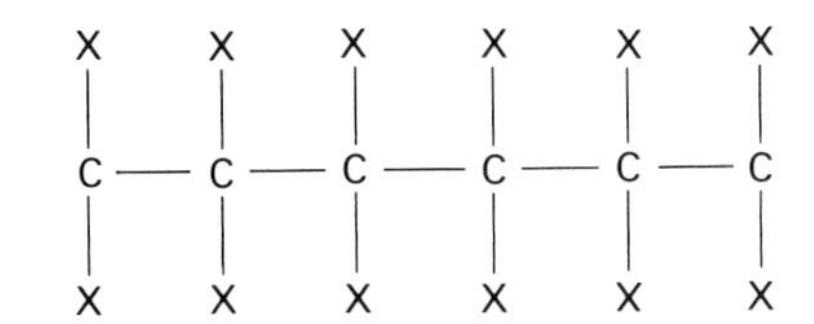

Fig. 7.6 Linear molecular structure

Alternatively, however, a *branched molecular structure* may be formed, as shown in Fig. 7.7.

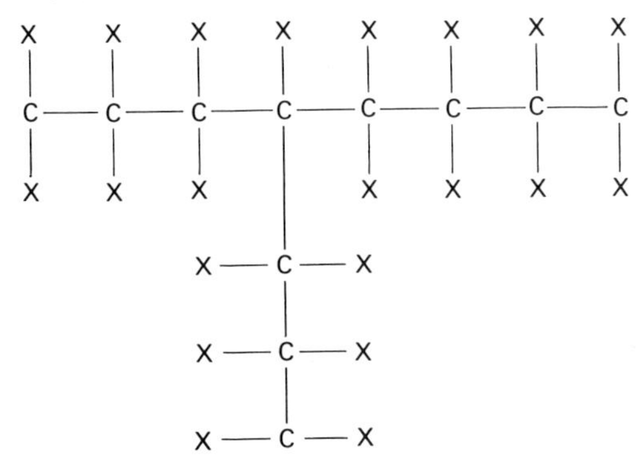

Fig. 7.7 Branched molecular structure

A more complex molecular structure may be formed by the linking together of these branches to form what is termed a *cross-linked molecular structure*, as shown in Fig. 7.8.

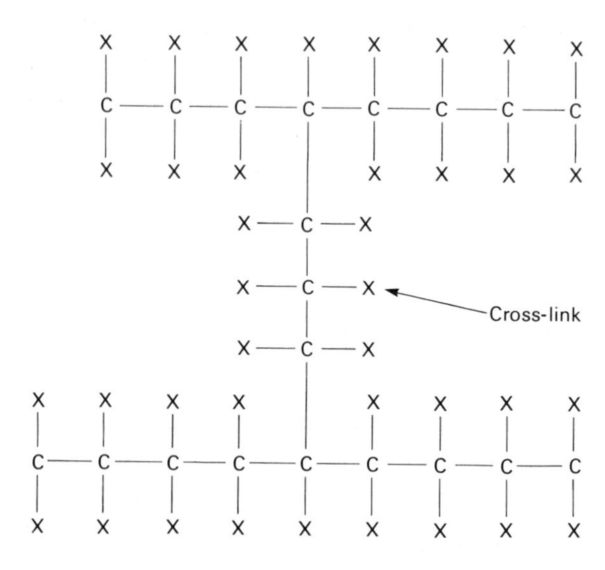

Fig. 7.8 Cross-linked molecular structure

The mechanical properties of polymers, like those of metallic materials, depend greatly on the molecular structure of the material. Generally, linear polymers will be very flexible, the linear chains readily sliding past each other when stress is applied to the material. A branched polymer, on the other hand, has a greater resistance to distortion, giving less flexibility but a corresponding increase in strength. Cross-linking of the molecular chains will still further increase the strength and rigidity of the polymer. An important example of the effect of cross-linking is the vulcanisation process for rubber, in which sulphur atoms enable cross-links to occur in the structure, thus increasing the hardness of the rubber. The resulting hardness may be varied by controlling the degree of cross-linking induced in the rubber by the sulphur atoms.

THERMOSETTING AND THERMOPLASTIC MATERIALS

Polymers may be classified as either *thermosetting* or *thermoplastic*. A thermosetting material becomes plastic when heated but quickly sets hard. The heating of the polymer causes extensive cross-linking of the molecular structure. The process is not reversible and the result is a hard brittle material. Phenolic and epoxide resins are examples of thermosetting materials.

In thermoplastic materials, the molecular structure is either linear or lightly branched and the material becomes more flexible when heated. This change is reversible and may be repeated any number of times, provided the temperature is not so high as to decompose the material. Such a thermoplastic material may be easily moulded if it is heated to a suitable temperature. Examples of thermoplastic polymers are polymethyl methacrylate (Perspex), polyethylene and PVC.

CRYSTALLINE AND AMORPHOUS POLYMERS

We saw in Chapter 1 that the properties of metallic materials are strongly influenced by the arrangement of the atoms, and by the type of unit cell present. In polymeric materials also, there may be some degree of atomic order present, so that some parts of the material may be called crystalline. In other parts the atomic arrangement will be random, and this is termed an *amorphous* (i.e. shapeless) structure. In any given polymer the structure is very unlikely to be either completely crystalline or completely amorphous, but materials are classified as either crystalline or amorphous according to the nature of the atomic arrangement in the *bulk* of the material. The simplest way in which some degree of crystallinity may be present is by the lining up of parts of different linear chain molecules to give some localised order to the arrangement. This is illustrated for polyethylene in Fig. 7.9.

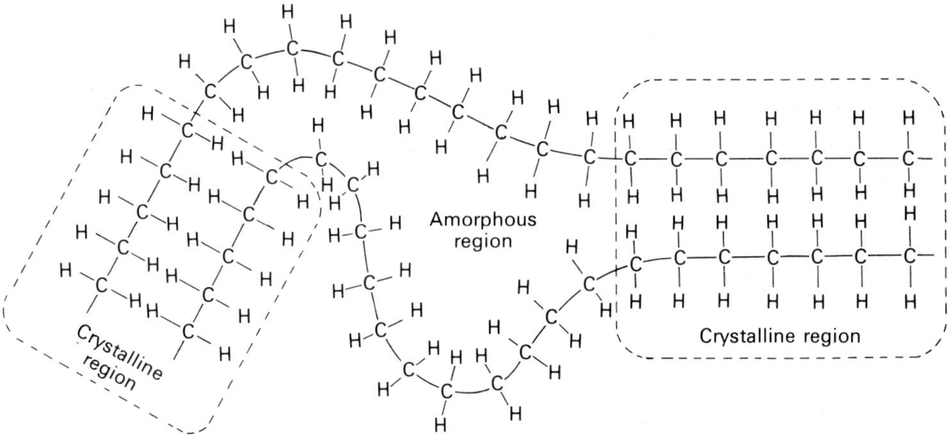

Fig. 7.9 Crystalline and amorphous regions in polyethylene

Another way in which crystalline regions may appear is by the folding of individual linear chains to give regions of atomic order, as in Fig. 7.10.

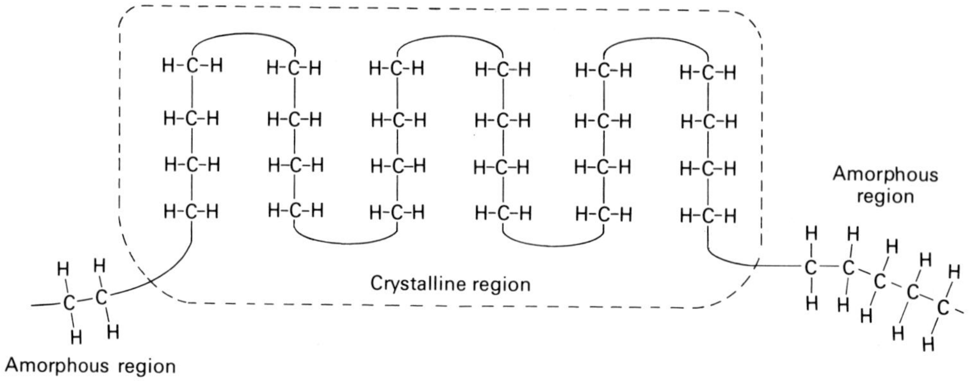

Amorphous region

Crystalline region

Amorphous region

Fig. 7.10 Crystalline region resulting from the folding of a single molecular chain

An amorphous material is generally brittle at low temperatures but tougher at higher temperatures. Crystalline materials, on the other hand, tend to be tough at low temperatures and the material may have a higher tensile strength. A crystalline polymer tends to have a higher density, because of the closer atomic packing. Materials such as polyethylene can exist either in a mainly amorphous form or in a mainly crystalline form – hence we have low-density or high-density polyethylene.

Stress–strain curves for polymer materials show considerable variations, due to the influence of the following factors:

(i) the nature of the polymer, i.e. whether the polymer is mainly amorphous or mainly crystalline, and whether extensive molecular cross-linking is present;
(ii) the temperature;
(iii) the rate at which the material is strained.

In some cases the polymer shows a stress–strain curve similar to that of a brittle metal (Fig. 7.11), whereas in other cases the curve may be of the type shown in Fig. 7.12, which characterises a tougher material. A given polymer may show either of these curves, depending on the temperature and the strain rate. The cold-drawing effect illustrated in Fig. 7.12 can produce the same result in polymer materials that cold working does in metals: a significant increase in tensile strength.

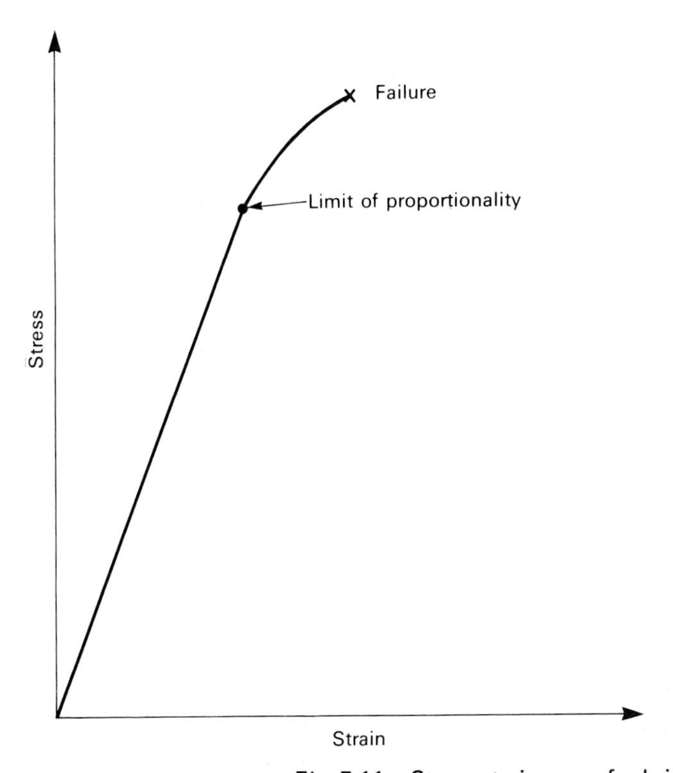

Fig. 7.11 Stress-strain curve for brittle polymer

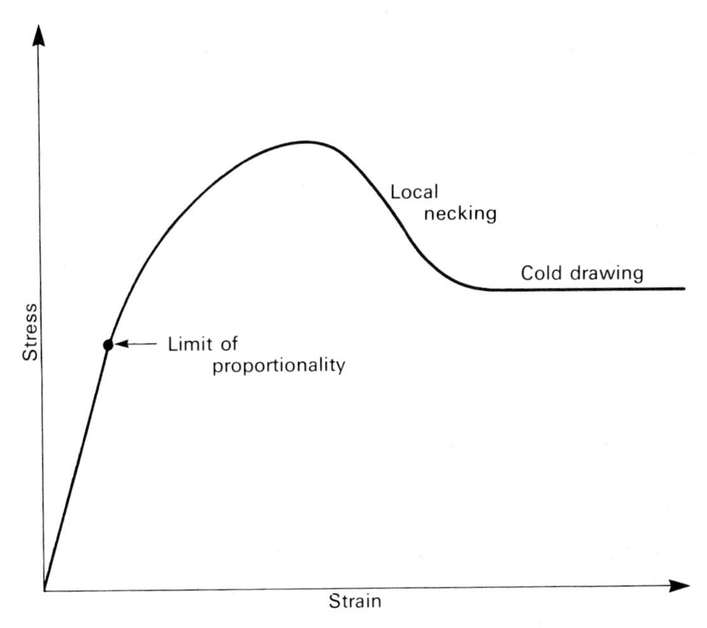

Fig. 7.12 Stress-strain curve for polymer exhibiting cold drawing (tough polymer)

PROPERTIES OF SOME COMMON POLYMER MATERIALS

Polyethylene

This is a thermoplastic material. Two types are obtainable: a low-density type, and a high-density type. The density obtained depends upon the method of polymerisation from the ethylene monomer. Low-density polyethylene has a lower degree of crystallinity, resulting in lower values of tensile strength, Young's modulus, percentage elongation at break, and softening temperature. The high-density material has a higher degree of crystallinity with correspondingly higher values of these properties. The softening temperature determines how easily the material can be formed and the maximum service temperature acceptable. The softening temperature of low-density material is less than 100 °C, while that of high-density material is above 100 °C. The higher softening temperature of the high-density material allows its use in medical applications, since sterilisation at 100 °C may be carried out. The material is sometimes prone to stress cracking in certain chemical environments but suitable additives may be used to overcome this problem.

Polyethylene has good resistance to most chemicals and is used extensively in the form of sheets and bags for packaging. Other applications include water pipes, kitchen ware, and electrical insulation for cables, especially for very high radio frequencies. Typical properties of polyethylene are shown in Table 7.1.

TABLE 7.1 *Properties of polyethylene*

	Low-density	High-density
Density (kg/m^3)	920	950
Tensile strength (N/mm^2)	14	30
Young's modulus (N/mm^2)	200	800
Percentage elongation at break	100–650	50–850
Softening temperature (°C)	95	110

Polyvinyl chloride (PVC)

This is an amorphous thermoplastic material. The vinyl chloride monomer is derived from ethylene, and polymerisation results in a linear polymer as shown in Fig. 7.5, on p. 131. The presence of the chlorine atoms tends to stiffen the polymer, and PVC is a rigid material. The material may, however, be softened by the addition of a 'plasticiser' to the polymer. The flexibility of the PVC may be varied by adjusting the amount of plasticiser added and thus a wide range of properties may be obtained. Other materials are usually added to reduce the cost of the material, to provide colouring or to confer specific properties such as improved resistance to certain chemicals. Most PVC components are produced from plasticised material but non-plasticised high-impact PVC is used for certain applications. Typical properties of the two types of PVC are given in Table 7.2.

TABLE 7.2 *Properties of PVC*

	Rigid (unplasticised)	Plasticised
Density (kg/m^3)	1400	1300
Tensile strength (N/mm^2)	56	35
Percentage elongation at break	20	220

Plasticised PVC is used for upholstery, conveyor belts, pipes, cable insulation and rainwear. Unplasticised PVC is used for cold water pipes, refrigerator linings and chemical plant.

Polypropylene

Polypropylene is a thermoplastic material which is usually crystalline in structure and is obtained by the polymerisation of propylene.

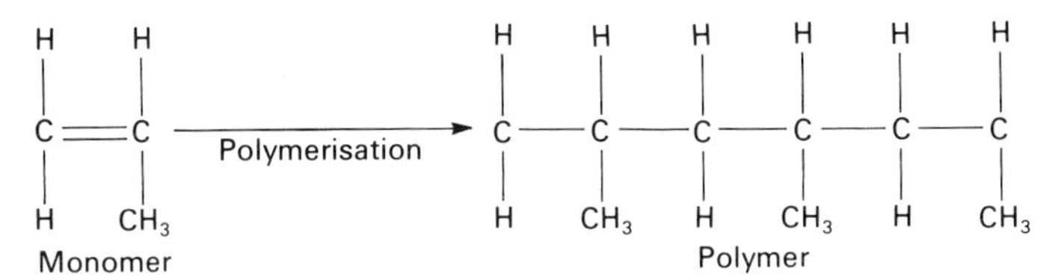

Fig. 7.13 Formation of polypropylene

This polymer is stronger than polyethylene and has a higher softening temperature. The material has a low density and retains its tensile strength up to moderately high temperatures (100 °C). The main applications of polypropylene are for packaging, electrical insulation, bottles, automobile components, refrigerator linings and washing machine parts. Typical properties are shown in Table 7.3.

TABLE 7.3 *Properties of polypropylene*

Density (kg/m^3)	900
Tensile strength (N/mm^2)	40
Percentage elongation at break	300

Polystyrene

This is an amorphous thermoplastic material produced by the polymerisation of the styrene monomer.

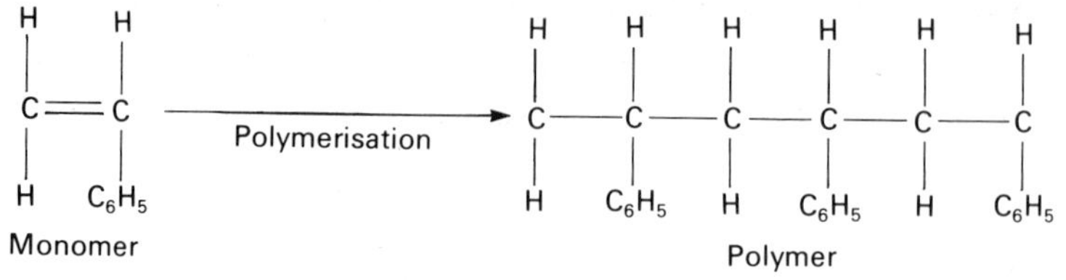

H H
C=C
H C_6H_5
Monomer

Polymerisation

H H H H H H
C—C—C—C—C—C
H C_6H_5 H C_6H_5 H C_6H_5
Polymer

Fig. 7.14 Formation of polystyrene

The material tends to be brittle, and additives are used to reduce this and to improve its impact resistance. Polystyrene suitably blended with other materials is used to produce mouldings for use in washing machines and similar household appliances. Its resistance to most foods leads to its use for food containers. One of the most important applications is the use of the material suitably 'foamed' to form cushioning mouldings for packaging fragile goods. The retention of properties at low temperatures leads to its use for refrigerator and deep-freeze casings.

A wide range of more complex polymers has been developed from styrene: the ABS polymers, which are produced by forming copolymers of the three polymers *acrylonitrile, butadiene* and *styrene* – hence the abbreviation ABS. They have a high resistance to many chemicals, high impact resistance and retain their properties at higher temperatures (above 100 °C). Typical applications for these rather more expensive polymers are battery cases and computer cases. Table 7.4 shows typical properties of polystyrene and ABS.

TABLE 7.4 *Properties of polystyrene and ABS*

	Polystyrene	**ABS**
Density (kg/m^3)	1100	1100
Tensile strength (N/mm^2)	35–60	17–58
Percentage elongation at break	2	80
Maximum service temperature (°C)	70	110

Polytetrafluoroethylene (PTFE)

This is a crystalline thermoplastic formed by polymerisation of the monomer tetrafluoroethylene, as shown in Fig. 7.4 on p. 130. This polymer is chemically inert, has a very low coefficient of friction, and is thermally stable up to temperatures of the order of 300 °C. Because the material does not readily flow, even at high temperatures, components are formed by cold or hot compacting followed by sintering (heating the component at a high temperature).

PTFE is used as an electrical insulating material, and for low-stress bearings, chemically resistant linings for chemical processing plant and non-stick coatings for frying pans, etc. The properties of PTFE are given in Table 7.5.

TABLE 7.5 *Properties of PTFE*

Density (kg/m^3)	2200
Tensile strength (N/mm^2)	25
Percentage elongation at break	250
Coefficient of friction	<0.1

Polyamides (nylons)

These are thermoplastic crystalline materials with linear molecular chains. There is a wide range of polymers, but the most important types are known as nylon 6, nylon 66, nylon 610 and nylon 11, the figures indicating the number of carbon atoms in the substances reacting to form the polymer. The main characteristics of this group of polymers are their high strength, high softening temperature and good chemical resistance. One disadvantage of the nylon group is a tendency to absorb moisture, with a resulting decrease in tensile strength. The high melting point and softening temperature of the nylons allows them to be used at temperatures in excess of 100°C. The main uses of the materials are as fibres for the textile industry and in the form of moulded gears, bearings and containers. The possibility of moisture absorption reduces their effectiveness as electrical insulating materials. While they offer good resistance to many organic chemicals they are attacked by many acids. Table 7.6 shows properties for the different nylons.

TABLE 7.6 *Properties of nylons*

	Nylon 6	Nylon 66	Nylon 610	Nylon 11
Density (kg/m^3)	1100	1100	1100	1100
Tensile strength (N/mm^2)	80	80	60	50
Percentage elongation at break	100–300	60–300	80–250	70–300

Elastomers

As the name implies, these are elastic materials which, although distorting extensively when stressed, return to their original shape immediately on removal of the applied stress. Natural rubber is the best known example of such a material, but synthetic rubbers are now used much more extensively. Natural rubber is polyisoprene, and the nature of the polymer chain is such that cross-linking readily occurs between the polymer chains. This cross-linking is known as *vulcanisation* and is brought about in a controlled manner by heating the rubber while sulphur is applied to it. The sulphur is absorbed by the rubber, the sulphur atoms acting as the cross-links between the molecular chains. The principle is illustrated in Fig. 7.15.

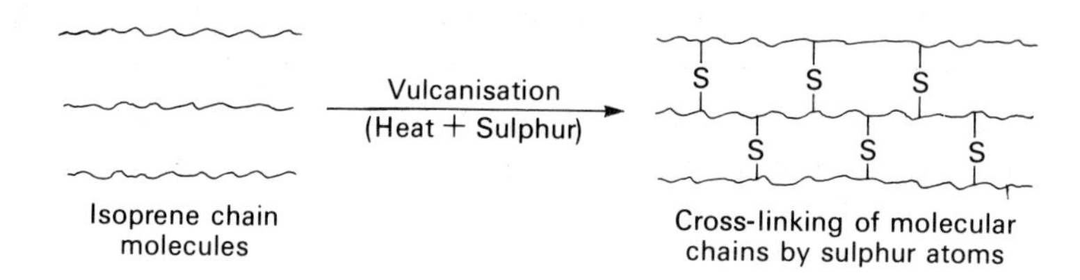

Fig. 7.15 Vulcanisation of rubber

The curing or vulcanisation temperature is of the order of 150 °C. The greater the amount of sulphur added, the more extensive will be the cross-linking, with a corresponding increase in the rigidity of the rubber. When large amounts of sulphur are added, complete vulcanisation may occur and the material then becomes fully rigid (*ebonite*).

There is a wide range of synthetic rubbers based on other polymers and produced from the distillation of oil. The synthetic rubbers known as *Neoprene* and *nitrile* rubber have better wear and temperature resistance than natural rubber. *Silicone rubbers* retain their flexibility over a very wide temperature range (−90 °C to 250 °C is typical) so they can even be used as flexible moulds for thermoplastics.

Natural rubbers are attacked by most greases and oils, whereas some synthetic rubbers have good resistance to these substances.

Thermosetting resins

Thermosetting resins are used extensively in the form of mouldings and as adhesives. One of the earliest resins produced was the *phenolic* type (Bakelite) in which fillers are used to reduce the cost and to provide specific properties. Typical applications are electrical mouldings, switches and laminates.

A second group of thermosets is known as the *polyester resins*. In these, polymerisation results in an extensive network type of molecular structure. Curing of the resin may either take place at room temperature or be accelerated by heating at temperatures of the order of 100 °C. This resin is used to form laminate materials, using glass fibres as the reinforcing and stress-carrying component of the laminate.

Epoxide resins have found extensive use as adhesives in recent years under the trade name Araldite. The resin is mixed with a curing agent and the resulting chemical reaction converts the liquid resin and liquid curing agent (hardener) into a rigid solid material. The resin is used for sealing or 'potting' electrical components and also, in combination with carbon fibres, to form a very strong and stiff composite material for aerospace components, fishing rods, and (recently) golf clubs. Epoxide resins combined with suitable fillers (stone chippings, etc.) are used to form hard-wearing flooring compounds. The main advantages of these resins are that they are resistant to most chemicals and have high mechanical properties.

Stabilisers, plasticisers and fillers

Polymer materials may be degraded by the action of chemicals, by ozone in the atmosphere, or by exposure to visible light or higher-energy radiation such as X-rays. In many cases, materials are added to the polymer during mixing or polymerisation to either reduce or eliminate this degradation, and these additives are known as *stabilisers*.

Other materials are sometimes added to improve the flexibility of the polymer, and such substances are known as *plasticisers*. In some cases two versions of the polymer are available: an unplasticised type or a plasticised type, to suit the properties required of the finished product.

Materials known as *fillers* are used for a variety of reasons. For example, the filler may be a low-cost material used to 'bulk out' the polymer and thus reduce the cost per unit volume without having an unacceptable effect on the required properties. Other fillers may be added to colour the polymer, or to alter its electrical conductivity – this may be done by adding metallic particles or powdered mica to the polymer. Magnetic rubbers may be produced by adding iron particles to the basic polymer.

SOME ADVANTAGES AND DISADVANTAGES OF POLYMER MATERIALS

ADVANTAGES

(i) Polymer materials are good electrical and thermal insulators.

(ii) Many polymer materials have very low densities (of the order of 1000 kg/m^3 compared with 2400 kg/m^3 for aluminium or 7900 kg/m^3 for steel).

(iii) They may be easily fabricated by extrusion, by casting techniques or by laminating.
(iv) They are resistant to many of the corrosive agents which adversely affect metallic materials.
(v) In certain applications where high production quantities are involved (e.g. kitchen ware) components may be produced at a lower unit cost than the equivalent metallic component would be.
(vi) By the addition of suitable fillers a wide range of colours and specific properties may be produced.
(vii) Many polymer materials are non-toxic.

DISADVANTAGES

(i) Generally these materials have low strength. Due to their low density, however, the specific strength (tensile strength/density) may compare favourably with certain metals at room temperature.
(ii) They have a poor temperature resistance; maximum service temperatures are below 100 °C for many polymer materials and very rarely exceed 200 °C.
(iii) They may degrade due to the action of atmospheric ozone and sunlight.
(iv) Under load, and particularly at moderate temperatures, extensive 'creep' of polymer materials occurs. This only becomes a problem in metals at much higher temperatures.
(v) The cost of producing complex shapes in small production quantities is often prohibitive in comparison with metals.

SUMMARY

- *Polymers* are materials built up from a series of smaller units (*monomers*) to form long chain-like molecules which may vary in size from a few hundred of the basic unit to perhaps hundreds of thousands. The most common polymers have a 'backbone' of carbon atoms. The carbon atom has a valency of four, meaning that it can easily bond to two atoms with a valency of one and to another carbon atom on either side. The monomer may consist of a pair of carbon atoms, bonded to each other and each to three atoms of valency one – this forms a *saturated* monomer. Alternatively the carbon atoms may have a double bond to each other, and be bonded each to two atoms of valency one – this is an *unsaturated* monomer. An unsaturated monomer is easier to polymerise, because the double bond is easier to break than a single bond.

- Polymerisation tends to produce linear molecules in which the chain-like molecular structure may wind around an axis, like a spring. This gives a very flexible material, as the 'springs' can extend or contract and the molecular chains slide past each other fairly easily. It also enables the material to be moulded easily – raising the temperature to about 100°C relaxes the bonds between molecules allowing them to take up new positions. The material becomes rigid again when cold, and the cycle of heat to relax, cool to reset can be repeated almost indefinitely. Such materials are called *thermoplastics*.

- Branches may form on a linear polymer where a carbon atom, instead of being bonded to two atoms of valency one, carries only one such atom, and in place of the other is bonded to the carbon atom of another chain. This is called a *branched molecular* structure; it results in a less flexible but stronger material. If a branch from one chain is similarly joined at the other end of the branch to a similar chain, we have a *cross-linked molecular structure*, which will be even stronger and more rigid.

- In general, a polymer may be lightly branched and still remain a thermoplastic, but extensive cross-linking produces a rigid, brittle material. Such materials, phenolic and epoxide resins for example, are *thermosetting*, not thermoplastic. That is, they are plastic when heated but quickly set hard as extensive branching and cross-linking occurs. This process is not reversible. The most commonly known thermoset is *Bakelite*.

- *Elastomers* are materials such as rubber, which stretch considerably under load, but deform *elastically*; that is, they return to their original dimensions when the load is removed. The original elastomer was natural rubber. This is *cured* or *vulcanised* by treating with sulphur at about 150°C. The sulphur is absorbed and its atoms form cross-links between the molecular chains of the rubber. The hardness of the rubber depends on the amount of cross-linking, and hence on the amount of sulphur absorbed.

- Natural rubber has now largely been superseded by synthetic rubbers, which have better resistance to wear, to extremes of temperature and to attack by lubricants than natural rubber has.

- In general, the arrangement of the molecular chains forming a polymer tends to be random and such a molecular structure is called *amorphous*. However, it is possible for the units which make up the molecular chains to align themselves into parallel rows, and such a molecular structure is called *crystalline*. This alignment can either be between parallel chains of

TABLE 7.7 *Properties of some common polymeric materials*

Material	Category	Type	Density (kg/m^3)	Tensile strength (N/mm^2)	Young's modulus (N/mm^2)	Elongation (%)	Softening temperature (°C)	Typical applications
Polyethylene (polythene)	Thermoplastic Semi-crystalline	Low-density	920	14	200	100–650	95	Water pipes, kitchen-ware, insulation for UHF cables, sterilisable equipment (high density only)
		High-density	950	30	800	50–850	110	
Polyvinyl chloride (PVC)	Thermoplastic Amorphous	Rigid	1400	56	★	20		Cold water pipes, refrigerator linings, chemical plant
		Plasticised	1300	35	★	220		Upholstery, cable insulation, rainwear, baby pants
Polypropylene	Thermoplastic Crystalline		900	40	★	300	100	Packaging, electrical insulation, bottles, car components, washing machine parts
Polystyrene	Thermoplastic Amorphous	Polystyrene	1100	35–60	★	2	70	*Blended*: household mouldings. *Foamed*: packaging, thermal insulation
		ABS	1100	17–58	★	80	110	Battery cases, luggage, parts for electro-plating

TABLE 7.7 *Continued*

Material	Category	Type	Density (kg/m^3)	Tensile strength (N/mm^2)	Young's modulus (N/mm^2)	Elongation (%)	Softening temperature (°C)	Typical applications
Polytetra-fluoroethylene (PTFE)	Thermoplastic Crystalline		2200	25	★	250	327	Non-stick frying pans, low-friction bearings, electrical insulation, chemical plant
Polyamides (nylons)	Thermoplastic Crystalline	Nylon 6	1100	80	★	100–300	> 100	Fibres for textiles, gear wheels, bearings, containers. Unreliable for electrical insulation – absorbs moisture
		Nylon 66	1100	80	★	60–300		
		Nylon 610	1100	60	★	80–250		
		Nylon 11	1100	50	★	70–300		
Phenolic resins (e.g. Bakelite)	Thermosetting				★			Electrical mouldings, switches, laminates
Polyester resins	Thermosetting				★			Combined with styrene or glass fibre for reinforced polymer materials
Epoxide resins	Thermosetting				★			Adhesives, 'potting' electrical components, reinforced with carbon fibres for composite materials, used with fillers for flooring

★Values of Young's modulus vary considerably, as they depend on the condition of the polymer

adjacent molecules, or it can result from the folding of a single molecule into parallel lengths which lie alongside each other. Both crystalline and amorphous regions can exist within the same polymer.

- Crystalline polymers have a higher density, higher tensile strength and are less subject to low-temperature brittleness than amorphous polymers, but amorphous polymers tend to be tougher at higher temperatures.
- The form of stress–strain curve produced by a tensile test on a polymer depends on whether the material is amorphous or crystalline, whether there is extensive cross-linking, the temperature of the material and the rate at which the material is strained. Brittle polymers tend to have the same shape of stress–strain curve as brittle metals, and ductile polymers as ductile metals, but both forms of stress–strain curve can be produced from the same material by altering the temperature of the material, or the rate at which it is strained. Cold drawing a polymer can produce the same effect as cold working a metal: increased tensile strength.
- *Stabilisers* are added to polymers to prevent or delay the deterioration caused by ozone in the atmosphere or by exposure to sunlight.
- *Plasticisers* are added to improve the flexibility of a polymer such as PVC.
- *Fillers* are added to dilute the polymer with less expensive material, to colour it, or to alter properties such as electrical conductivity.
- The main properties of some common polymeric materials are given in Table 7.7 on pp.144–5.

EXERCISE 7

1) Why is the ethylene molecule more reactive than the ethane molecule?

2) With the aid of a diagram show the difference between a linear molecular structure and a branched molecular structure.

3) What is the difference in behaviour between a thermosetting and a thermoplastic material on heating?

4) Sketch a stress–strain curve for a polymer which exhibits the 'cold drawing' effect.

5) How may crystalline regions be formed in a polymer structure?

6) What effect does the degree of crystallinity have on the mechanical properties of a polymer?

7) Why is plasticised PVC used for cable insulation while unplasticised PVC is used for refrigerator cases?

8) State the properties of the thermoplastic material known as PTFE (polytetrafluoroethylene).

9) Describe the process known as *vulcanisation*.

10) Why are (a) filler materials, (b) stabilising materials, added to many polymers?

8 MACRO- AND MICRO-EXAMINATION OF METALLIC MATERIALS

MACRO-EXAMINATION

'Macro' comes from a Greek word meaning 'large', and macro-examination means the examination of specimens either with the naked eye or at low magnifications, using a magnifying glass or a microscope. A suitable but arbitrary distinction between macro- and micro-examination may be made by defining macro-examination as examination of a specimen at magnifications lower than ×10. Examination at low magnifications is important when studying fracture faces, or welded or brazed joints. It is also useful in the detection of surface corrosion or large-scale defects such as inclusions or porosity in castings. In many cases, for example in the examination of fracture faces, surface preparation is not required; but even where it is necessary, the surface does not usually need to be highly polished. Where sectioning of the component is required, it is very important that suitable sketches and photographs of the main features are recorded. This should include a sketch indicating the position and nature of the section to be cut from the specimen.

Normally the section is either cut by hand or by the use of a suitable off-cut machine, care being taken to ensure that the cutting process does not overheat the material. If the section is small enough it may be possible to mount it in a Bakelite type of material before it is polished. This procedure is explained more fully on the next page.

The section is ground using lubricated emery paper, and for macro-examination a final finish on a Grade O paper is usually sufficient. The specimen should then be examined in this unetched condition, and any significant features noted. A suitable etching reagent (*etchant*) is then applied to the surface of the section, usually by swabbing with cotton wool immersed in the reagent.

This is particularly useful in determining the structure and presence of defects in a weld section, or for the determination of grain flow in forgings. Since most etching reagents are either acids or strong alkalis, care should be taken in their use. The etching reagent shows up certain structural features such as changes in grain direction, large inclusions or blow holes, by attacking one form of surface material more deeply than another. The etching time may vary from a few seconds to several hours in the case of large forging sections. After being etched, the section is swilled with water, and then swabbed with methyl alcohol to promote rapid drying.

MICRO-EXAMINATION

The range of the optical microscope provides magnifications up to ×3000, although more usually magnifications up to ×1000 are used. The metallurgical microscope differs from the biological microscope in that normal incident light is made to reflect from the surface of the opaque specimen instead of passing through the specimen, as is usual for biological specimens. The principle of the standard metallurgical microscope is illustrated in Fig. 8.1. The important feature of this microscope is the semi-reflecting mirror in the form of a glass disc, which reflects the light beam down on to the specimen. The light reflected by the surface of the specimen then passes up through the mirror to the eyepiece lens.

The first stage in the preparation of a section for micro-examination is to photograph and/or sketch the position and nature of the section relative to the specimen as a whole. The section is cut as for macro-examination, either by hand sawing or by the use of a suitable off-cut machine, care being taken to avoid overheating the section. Where possible, the section is mounted in a hot-curing thermosetting resin such as Bakelite. Specialised presses are available for this purpose, with facilities for heating the resin to the appropriate cure temperature (usually about 150°C), cooling the encapsulated specimen by means of a water-jacket and finally ejecting it from the press. The advantages of using resin-mounted specimens are:

(i) manual handling of the specimens is easier; and
(ii) final polishing may be carried out on automatic polishing machines.

Grinding of the specimen is carried out by working through a series of four silicon carbide papers, commencing at a coarse 240 grade and finishing on a grade 600 paper. This is usually done by hand on a pre-grinder, in which the papers are clipped to a glass plate with running water continuously swilling

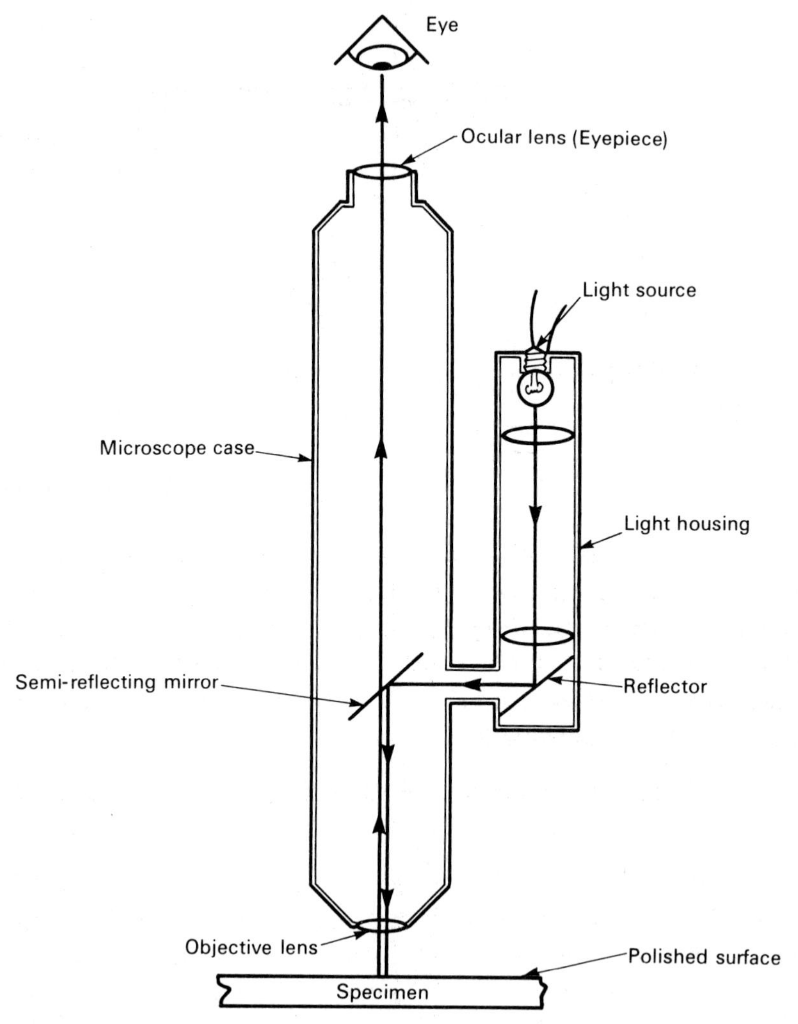

Fig. 8.1 Cross-section of a metallurgical microscope using normal incident light

the paper (see Fig. 8.2). The specimen is not transferred to the next grade of paper until the scratches from the previous grade of paper have been eliminated. The specimen is turned through 90° between papers. It is advisable to swill the specimen under the tap between papers, to avoid carrying a coarser grinding dust over to the next paper. After the final grind on the grade 600 paper, the specimen is ready for final polishing.

It is usual to carry out the final polishing operation on rotating discs, which often have a facility for automatic polishing, using a head in which the specimen mount is located. The polishing pad, which may be of a proprietary type or *selvyt* cloth, is impregnated with diamond paste suitably lubricated.

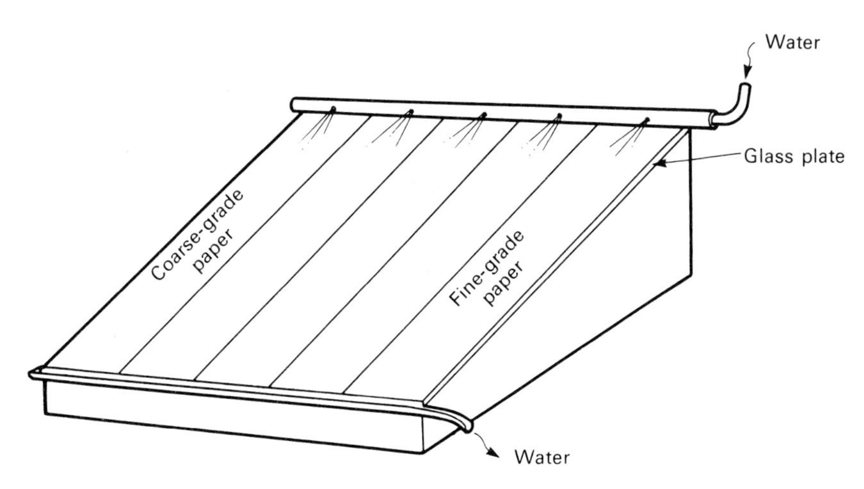

Fig. 8.2 Typical pre-grinder

The specimen is polished through three grades of diamond paste ranging from a 6 micron to a 1 micron or even a $\frac{1}{4}$ micron grade. The specimen should be washed between polishing stages to avoid contaminating a pad with coarser paste from the previous pad. For a high-quality finish or for photomicrography it is often desirable to carry out a final polish using a fine magnesia or fine alumina paste.

After a final wash the specimen is dried, and examined at both low and high magnification in this unetched condition.

The specimen is then etched in a suitable reagent, either by immersion in a dish or by swabbing with cotton wool impregnated with the etchant. The etching should be done cautiously, because the surface will have to be repolished if it has been over-etched, whereas further etching can easily be carried out if it is found to be necessary when the specimen is examined under the microscope. After etching, the specimen is washed first in water, then in alcohol, and finally dried under a warm air blower. Typical etchants are shown in Table 8.1; further information on these may be obtained by reference to specialised publications.

After the specimen has been etched, it is examined over a range of magnifications. The effect of etching in bringing out structural detail is illustrated in Fig. 8.3. In Fig. 8.3(a), which represents a specimen after polishing but before etching, there will be uniformly perpendicular reflection of light from the specimen, and no structural detail will be observed. Fig. 8.3(b) shows the effect of an etchant which has corroded the grain boundary regions only, so that the scatter of light from these regions will reveal grain boundary detail. Fig. 8.3(c) illustrates the effect of an etchant which, in addition to corroding

TABLE 8.1 *Typical etching agents*

Material	Etchant
Steel	(i) 2% 'Nital': 2 ml concentrated nitric acid in 100 ml alcohol. Standard etch for steels. (ii) 5% 'Nital': 5 ml concentrated nitric acid in 100 ml alcohol. Suitable for macro-etching.
Stainless steels	(i) 10% solution hydrochloric acid in alcohol. (ii) 10% by volume hydrochloric acid, 10% by weight ammonium persulphate in water.
Some alloy steels	(i) 2% 'Nital'. (ii) Acid ferric chloride etchant: 5 g ferric chloride, 50 ml hydrochloric acid to 100 ml water.
Aluminium alloys	(i) Mixed acid etchant: 95 ml water, 1.5 ml hydrochloric acid, 2.5 ml nitric acid, 0.5 ml hydrofluoric acid. (ii) 5% sodium hydroxide in water.
Copper alloys	(i) 10 g ferric chloride, 30 ml hydrochloric acid, 120 ml water. (ii) 10 g ammonium persulphate in water. The rate of etching may be increased by the addition of up to 30 ml ammonium hydroxide.
Cast iron	(i) 2% 'Nital'. (ii) 5% 'Nital'.
Zinc and its alloys	(i) 2% 'Nital'. (ii) 10% sodium hydroxide in water.
Magnesium and its alloys	(i) 2% 'Nital'.

Etchants should be handled with care. In the event of splashing on any part of the body, wash liberally with water.

grain boundary regions, has corroded grains to varying degrees. In this case, light reflection will vary for different grains, so that some grains will appear lighter and others darker. Fig. 8.3(d) illustrates the effect of etching a two-phase structure such as pearlite in steel, and shows the resulting lamellar view through the microscope. The photomicrographs shown in Fig. 1.2 (p. 4) and Fig. 2.17 (p. 33) are examples corresponding to Figs. 8.3(b) and 8.3(d) respectively.

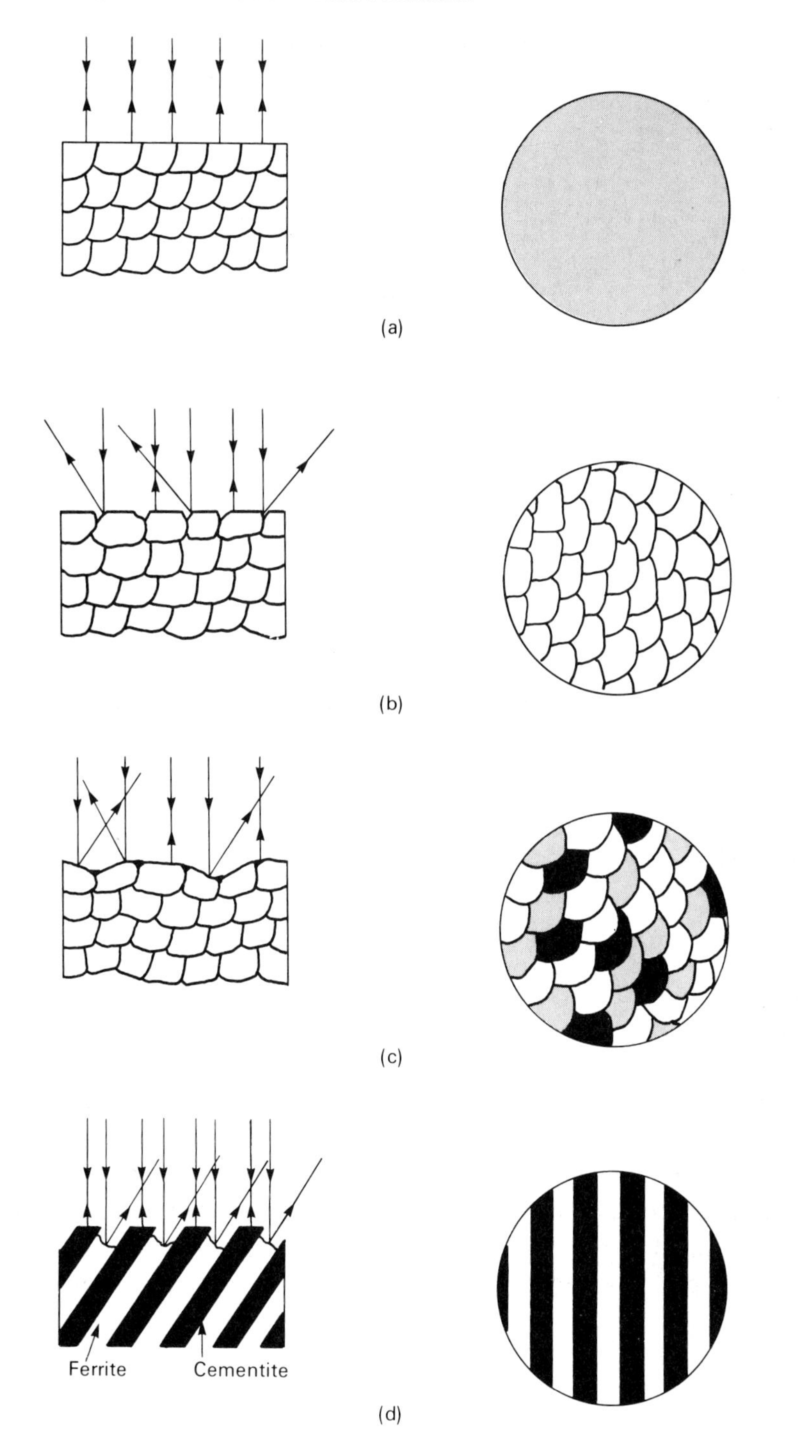

Fig. 8.3 Etching effects

REVISION CROSSWORDS

We hope these may entertain, as well as help you revise. As far as possible they are made up of words or abbreviations defined in this book, but of course we have had to use a few everyday words as well, to enable us to complete each puzzle.

As you can see, they start small – for beginners – and work up. Solutions on p. 169.

1

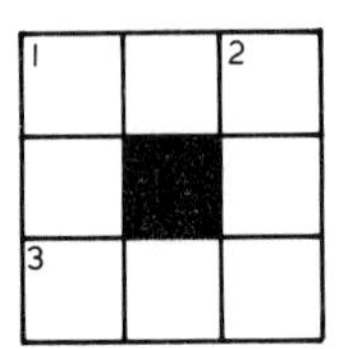

Across

1 Final inscription for one who handles cyanide carelessly? (1, 1, 1)
3 Unit cell of iron above 1400°C. (1, 1, 1)

Down

1 Feature of a forging which might be most likely to suffer from overheating? (3)
2 Thermoplastic amorphous polymer used for cable insulation. (1, 1, 1)

2

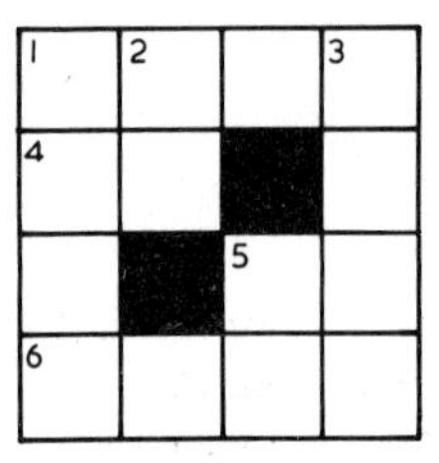

Across

1 Do this when preparing a section for the microscope. (4)
4 Protactinium, Dad! (2)
5 Toothpaste or strontium? (2)
6 Material used to form castings? (4)

Down

1 You might find it marked on 'silver' spoons. (1, 1, 1, 1)
2 Thanks for tantalum! (2)
3 Having a high resistance to indentation. (4)
5 Tin. (2)

3

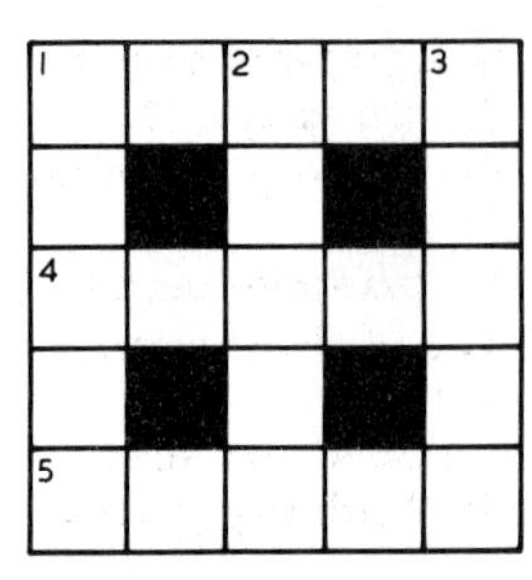

Across

1 A salty quench medium. (5)
4 The constituents of all matter. (5)
5 Silicon at one end, aluminium at the other, sulphur in middle – it's a kind of hemp. (5)

Down

1 An alloy of copper and zinc. (5)
2 Ferrous implements used by golfers. (5)
3 Artist's equipment. (5)

4

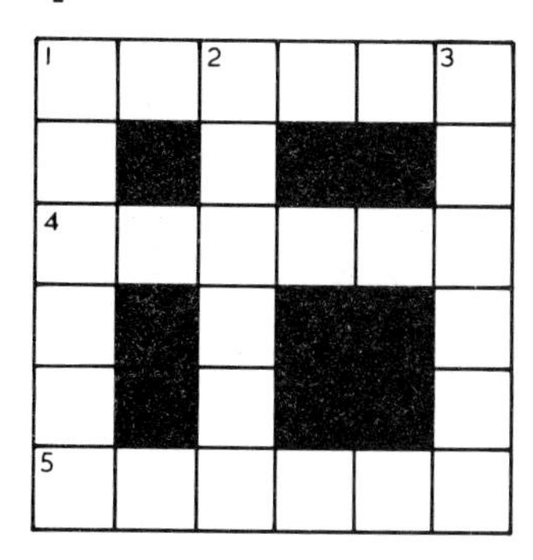

Across

1 Boundary line between two solid phases. (6)
4 Soften and toughen by heat treatment. (6)
5 Under a microscope, grain features which scatter reflected light appear ——. (6)

Down

1 Kept at constant temperature in a furnace. (6)
2 Capable of being represented on a graph by a straight line. (6)
3 Alloy of lead and tin. (6)

5

Across

1 Not a mixture or compound. (7)
4 It turns iron into steel. (6)
7 Machined surface on which a lathe saddle rests. (5)
8 Proposals to supply goods or services at a stated price. (7)

Down

1 Fluid used in preparing a specimen for the microscope. (7)
2 A form of completely vulcanised rubber. (7)
3 Titanium – it is upside down! (2)
5 General form of thermosetting material. (5)
6 Water might come from these in an induction-hardening process. (4)

6

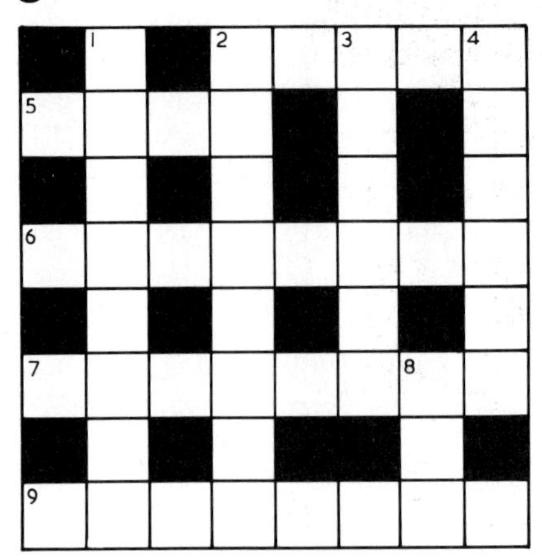

Across

2 Used to form internal spaces in castings. (5)
5 The point from which the reckoning begins in scales. (4)
6 It may be found in cast iron but not in steel. (8)
7 Boundary between completely liquid and partially solid phases. (8)
9 A crystalline shape forming in a cooling liquid metal. (8)

Down

1 Eutectoid mixture of ferrite and comentite. (8)
2 A substance having properties different from the properties of the substances from which it is made. (8)
3 Cause to remember. (6)
4 Load/area. (6)
8 Ultimate (abbreviation). (3)

7

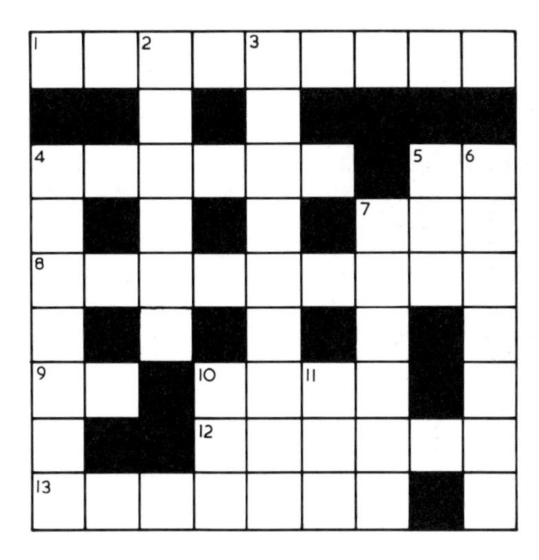

Across

1 Fe below 910 °C. (5–4)
4 A college site. (6)
5 Policeman (abbreviation). (2)
7 A conifer. (3)
8 The ability of a metal to be drawn out by tension. (9)
9 Indium is not out. (2)
10 A _____ gauge measures small displacements. (4)
12 '60% Cu, 40% Ni' as Archimedes might have put it? (6)
13 This type of single molecule joins others to make a polymer. (7)

Down

2 The hardened froth of certain glassy lavas. (6)
3 A light metal – good conductor of electricity. (9)
4 An element which, if present in a zinc-based diecasting alloy, makes it liable to intercrystalline corrosion. (7)
5 A small cavity in a metal surface. (3)
6 A solid whose atoms are arranged in a definite pattern. (7)
7 Material used to 'bulk out' a polymer or to alter its properties. (6)
10 To God (Latin). (3)
11 Exist in the plural. (3)

8

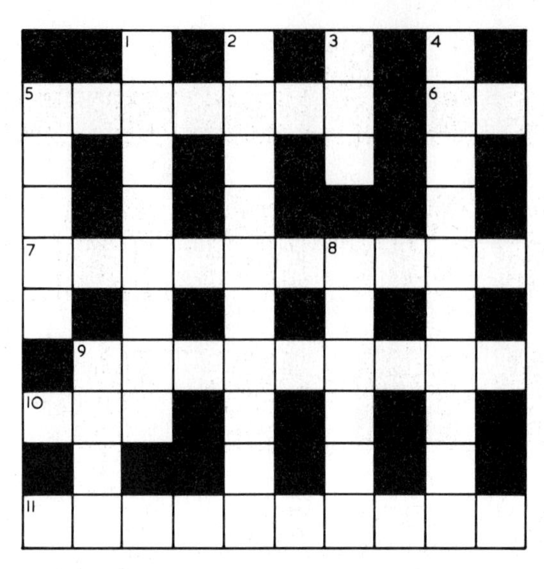

Across

5 A very dilute solution of carbon in body-centred cubic iron. (7)
6 This apparently masculine element is a light, inert gas. (2)
7 An item of optical equipment for a metallurgist. (10)
9 Fe above 1400 °C. (5–4)
10 The first colour radiated by a metal as its temperature is raised. (3)
11 A method of determining hardenability. (6, 4)

Down

1 In a ____ polymer structure, smaller chains of molecules project sideways from the main chains. (8)
2 A slight change of shape from that which was intended. (10)
3 Ferrous sulphide. (3)
4 Used to deoxidise copper. (10)
5 A ____ hardening process requires oxygen, acetylene and water. (5)
8 Person engaging the services of a professional such as an engineering consultant. (6)
9 Colloquial term for a procession with banners – or worse. (4)

9

Across

1 A unit cell of intermediate ductility. (1, 1, 1)

3 A policeman who is also an excellent conductor! (6)

7 A ____ material has an orderly arrangement of atoms. (11)

9 An insoluble solid produced by a reaction which occurs in solution. (11)

10 Alloy of iron and carbon with not more than 0.25% C. (4, 5)

11 Deoxidises steel; also makes chips! (2)

12 The lightest of the metals in common industrial use. (2)

13 The most severe of the common quench media. (5)

Down

2 The production of a hard surface layer on steel by heating in a carbonaceous medium. (11)

4 Amorphous thermoplastic ploymer, foamed to make cushioned packaging. (11)

5 The first German numeral. (3)

6 It is added to a polymer to prevent degradation. (10)

8 Point at which two or three solid constituents are simultaneously formed from another solid. (9)

10

Across

4 This component, usually made of an elastomer, keeps fluids in or out. (4)
7 This is often the primary consideration when deciding on a material for a component to be produced in quantity in a competitive industry. (4, 4)
9 A long way away. (3)
10 The second heat treatment which must be carried out after carburising. (4, 8)
11 An inert gas of atomic number 18. (2)
12 This surface hardening process does not shrink the material – in spite of its name! (12)
16 A piece of equipment used for the preparation of metallurgical specimens before they are polished. (3–7)
18 Removing oxygen, chemically. (8)
19 Large vessel with a tap from which tea or coffee is dispensed. (3)
21 A metallic element of atomic number 42. (2)
22 See 5 down.

Down

1 Girl, or former coin of India? (4)
2 The low-friction, high-melting-point polymer. (4)
3 The first heat treatment which must be carried out after carburising (4, 8)
5 and 22 A plot of temperature versus percentage composition, showing the temperatures at which various compositions of an alloy undergo phase changes when cooled slowly. (11, 7)
6 After severe cold working, the structure of a metal may consist of grains which are ____ and thin. (4)
8 A unit of weight for precious stones. (5)
10 Lettuce or trigonometrical ratio? (3)
13 Heat a steel to just above its A_3 temperature and then quench. (6)
14 A structure for drilling for oil. (3)
15 An individual crystal in a solidified metal. (5)
17 A mechanical device for gripping, or an animal? (3)
20 A horizontal line of entries in a table, or a noise? (3)

Appendix 1 THE CHEMICAL ELEMENTS

The names of elements that are significant in materials technology are printed in bold type. Some of the symbols are derived from Latin names (shown in parentheses).

Symbol	Name	Atomic number (number of electrons per atom)	Symbol	Name	Atomic number (number of electrons per atom)
Ac	Actinium	89	Eu	Europium	63
Ag	**Silver** (Argentum)	47	F	**Fluorine**	9
			Fe	**Iron** (Ferrum)	26
Al	**Aluminium**	13	Fm	Fermium	100
Am	Americium	95	Fr	Francium	87
Ar	Argon	18	Ga	Gallium	31
As	**Arsenic**	33	Gd	Gadolinium	64
At	Astatine	85	Ge	Germanium	32
Au	Gold (Aurum)	79	H	**Hydrogen**	1
B	**Boron**	5	Ha	Hahnium	105
Ba	**Barium**	56	He	Helium	2
Be	**Beryllium**	4	Hf	Hafnium	72
Bi	**Bismuth**	83	Hg	Mercury (Hydrargyrum)	80
Bk	Berkelium	97			
Br	Bromine	35	Ho	Holmium	67
C	**Carbon**	6	I	Iodine	53
Ca	**Calcium**	20	In	Indium	49
Cd	**Cadmium**	48	Ir	Iridium	77
Ce	Cerium	58	K	Potassium (Kalium)	19
Cf	Californium	98			
Cl	**Chlorine**	17	Kr	Krypton	36
Cm	Curium	96	La	Lanthanum	57
Co	**Cobalt**	27	Li	**Lithium**	3
Cr	**Chromium**	24	Lr	Lawrencium	103
Cs	Caesium	55	Lu	Lutetium	71
Cu	**Copper** (Cuprum)	29	Md	Mendelevium	101
			Mg	**Magnesium**	12
Dy	Dysoprosium	66	Mn	**Manganese**	25
Er	Erbium	68	Mo	**Molybdenum**	42
Es	Einsteinium	99			

(cont.)

Symbol	Name	Atomic number (number of electrons per atom)	Symbol	Name	Atomic number (number of electrons per atom)
N	**Nitrogen**	7	Ru	Ruthenium	44
Na	Sodium (Natrium)	11	S	**Sulphur**	16
			Sb	**Antimony** (Stibium)	51
Nb	**Niobium**	41			
Nd	Neodymium	60	Sc	Scandium	21
Ne	Neon	10	Se	Selenium	34
Ni	**Nickel**	28	Si	**Silicon**	14
No	Nobelium	102	Sm	Samarium	62
Np	Neptunium	93	Sn	**Tin** (Stannum)	50
O	**Oxygen**	8	Sr	Strontium	38
Os	Osmium	76	Ta	Tantalum	73
P	**Phosphorus**	15	Tb	Terbium	65
Pa	Protactinium	91	Tc	Technetium	43
Pb	**Lead** (Plumbum)	82	Te	Tellurium	52
			Th	Thorium	90
Pd	Palladium	46	Ti	**Titanium**	22
Pm	Promethium	61	Tl	Thallium	81
Po	Polonium	84	Tm	Thulium	69
Pr	Praseodymium	59	U	Uranium	92
Pt	Platinum	78	V	**Vanadium**	23
Pu	Plutonium	94	W	**Tungsten** (Wolfram)	74
Ra	Radium	88			
Rb	Rubidium	37	Xe	Xenon	54
Re	Rhenium	75	Y	Yttrium	39
Rf	Rutherfordium	104	Yb	Ytterbium	70
Rh	Rhodium	45	Zn	**Zinc**	30
Rn	Radon	86	Zr	**Zirconium**	40

Appendix 2 THE PERIODIC TABLE OF THE CHEMICAL ELEMENTS

The table overleaf gives the chemical symbol for the atom of each element, its atomic number (the number of electrons in the atom) and the distribution of those electrons between the various 'shells', from inner shell to outer shell. Thus, for example, the table shows that Iron (Fe) has 26 electrons, arranged 2 in the first (innermost) shell, 8 in the second, 14 in the third, and 2 in the fourth shell.

H 1 1																	He 2 2
Li 3 2, 1	Be 4 2, 2											B 5 2, 3	C 6 2, 4	N 7 2, 5	O 8 2, 6	F 9 2, 7	Ne 10 2, 8
Na 11 2, 8, 1	Mg 12 2, 8, 2											Al 13 2, 8, 3	Si 14 2, 8, 4	P 15 2, 8, 5	S 16 2, 8, 6	Cl 17 2, 8, 7	Ar 18 2, 8, 8
K 19 2, 8, 8, 1	Ca 20 2, 8, 8, 2	Sc 21 2, 8, 9, 2	Ti 22 2, 8, 10, 2	V 23 2, 8, 11, 2	Cr 24 2, 8, 13, 1	Mn 25 2, 8, 13, 2	Fe 26 2, 8, 14, 2	Co 27 2, 8, 15, 2	Ni 28 2, 8, 16, 2	Cu 29 2, 8, 18, 1	Zn 30 2, 8, 18, 2	Ga 31 2, 8, 18, 3	Ge 32 2, 8, 18, 4	As 33 2, 8, 18, 5	Se 34 2, 8, 18, 6	Br 35 2, 8, 18, 7	Kr 36 2, 8, 18, 8
Rb 37 2, 8, 18, 8, 1	Sr 38 2, 8, 18, 8, 2	Y 39 2, 8, 18, 9, 2	Zr 40 2, 8, 18, 10, 2	Nb 41 2, 8, 18, 12, 1	Mo 42 2, 8, 18, 13, 1	Tc 43 2, 8, 18, 14, 1	Ru 44 2, 8, 18, 15, 1	Rh 45 2, 8, 18, 16, 1	Pd 46 2, 8, 18, 18, 0	Ag 47 2, 8, 18, 18, 1	Cd 48 2, 8, 18, 18, 2	In 49 2, 8, 18, 18, 3	Sn 50 2, 8, 18, 18, 4	Sb 51 2, 8, 18, 18, 5	Te 52 2, 8, 18, 18, 6	I 53 2, 8, 18, 18, 7	Xe 54 2, 8, 18, 18, 8
Cs 55 2, 8, 18, 18, 8, 1	Ba 56 2, 8, 18, 18, 8, 2	La 57 2, 8, 18, 18, 9, 2	Hf 72 2, 8, 18, 32, 10, 2	Ta 73 2, 8, 18, 32, 11, 2	W 74 2, 8, 18, 32, 12, 2	Re 75 2, 8, 18, 32, 13, 2	Os 76 2, 8, 18, 32, 14, 2	Ir 77 2, 8, 18, 32, 17, 0	Pt 78 2, 8, 18, 32, 17, 1	Au 79 2, 8, 18, 32, 18, 1	Hg 80 2, 8, 18, 32, 18, 2	Tl 81 2, 8, 18, 32, 18, 3	Pb 82 2, 8, 18, 32, 18, 4	Bi 83 2, 8, 18, 32, 18, 5	Po 84 2, 8, 18, 32, 18, 6	At 85 2, 8, 18, 32, 18, 7	Rn 86 2, 8, 18 32, 18, 8
Fr 87 2, 8, 18, 32, 18, 8, 1	Ra 88 2, 8, 18, 32, 18, 8, 2	Ac 89 2, 8, 18, 32, 18, 9, 2															
				Ce 58 2, 8, 18, 20, 8, 2	Pr 59 2, 8, 18, 21, 8, 2	Nd 60 2, 8, 18, 22, 8, 2	Pm 61 2, 8, 18, 23, 8, 2	Sm 62 2, 8, 18, 24, 8, 2	Eu 63 2, 8, 18, 25, 8, 2	Gd 64 2, 8, 18, 25, 9, 2	Tb 65 2, 8, 18, 27, 8, 2	Dy 66 2, 8, 18, 28, 8, 2	Ho 67 2, 8, 18, 29, 8, 2	Er 68 2, 8, 18, 30, 8, 2	Tm 69 2, 8, 18, 31, 8, 2	Yb 70 2, 8, 18, 32, 8, 2	Lu 71 2, 8, 18, 32, 9, 2
				Th 90 2, 8, 18, 32, 18, 10, 2	Pa 91 2, 8, 18, 32, 20, 9, 2	U 92 2, 8, 18, 32, 21, 9, 2	Np 93 2, 8, 18, 32, 22, 9, 2	Pu 94 2, 8, 18, 32, 24, 8, 2	Am 95 2, 8, 18, 32, 25, 8, 2	Cm 96 2, 8, 18, 32, 25, 9, 2	Bk 97 2, 8, 18, 32, 26, 9, 2	Cf 98 2, 8, 18, 32, 28, 8, 2	Es 99 2, 8, 18, 32, 29, 8, 2	Fm 100 2, 8, 18, 32, 30, 8, 2	Md 101 2, 8, 18, 32, 31, 8, 2	No 102 2, 8, 18, 32, 32, 8, 2	Lr 103 2, 8, 18, 32, 32, 9, 2

SOLUTIONS

SELF-TEST QUESTIONS

Question 1

(a)

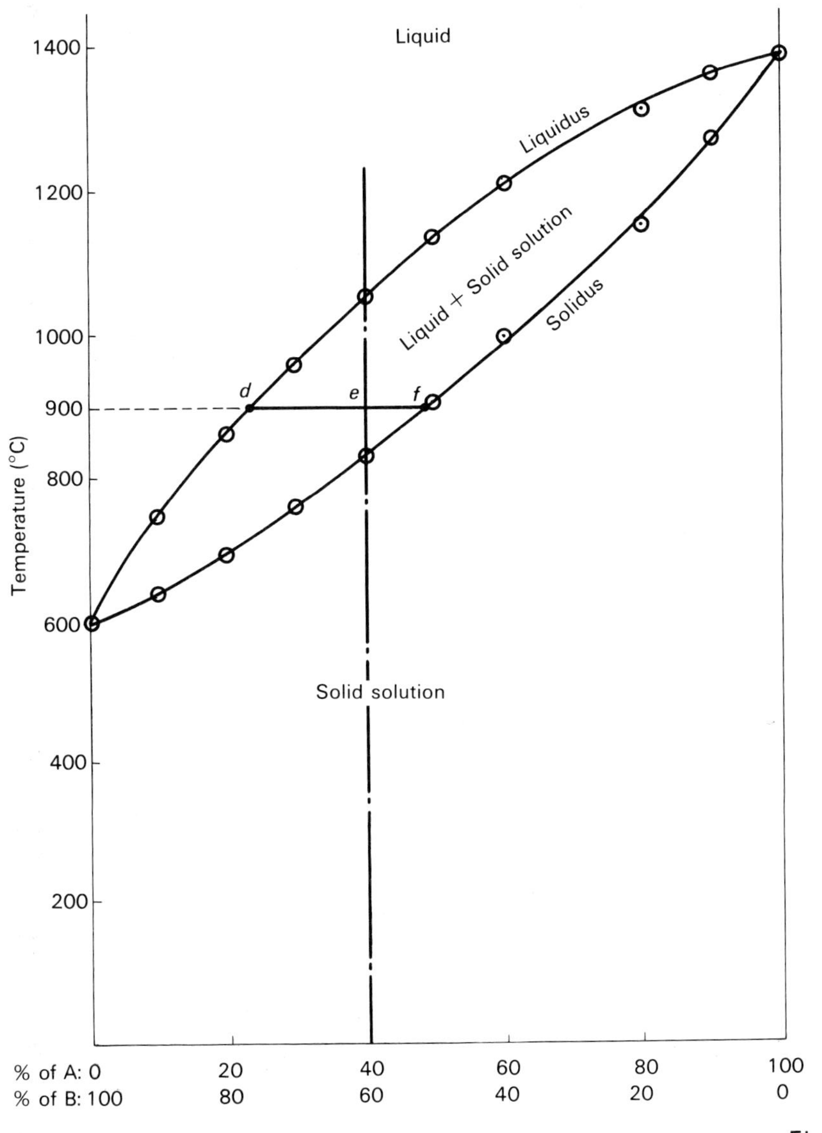

Fig. S.1

(b) From the plot of the equilibrium diagram, for the 40% A alloy:

(i) solidification commences at 1055°C;

(ii) solidification is completed at 830°C,

(iii) for this alloy, at 900°C, the composition of the liquid phase is given by point d, and is 23% A, 77% B, while the composition of the solid phase, given by point f, is 49% A, 51% B;

(iv) Ratio $\left(\frac{\text{Solid phase}}{\text{Liquid phase}}\right) = \frac{de}{ef} = \frac{17}{9}$

$= \frac{1.89}{1}$

Question 2

Equilibrium diagram:

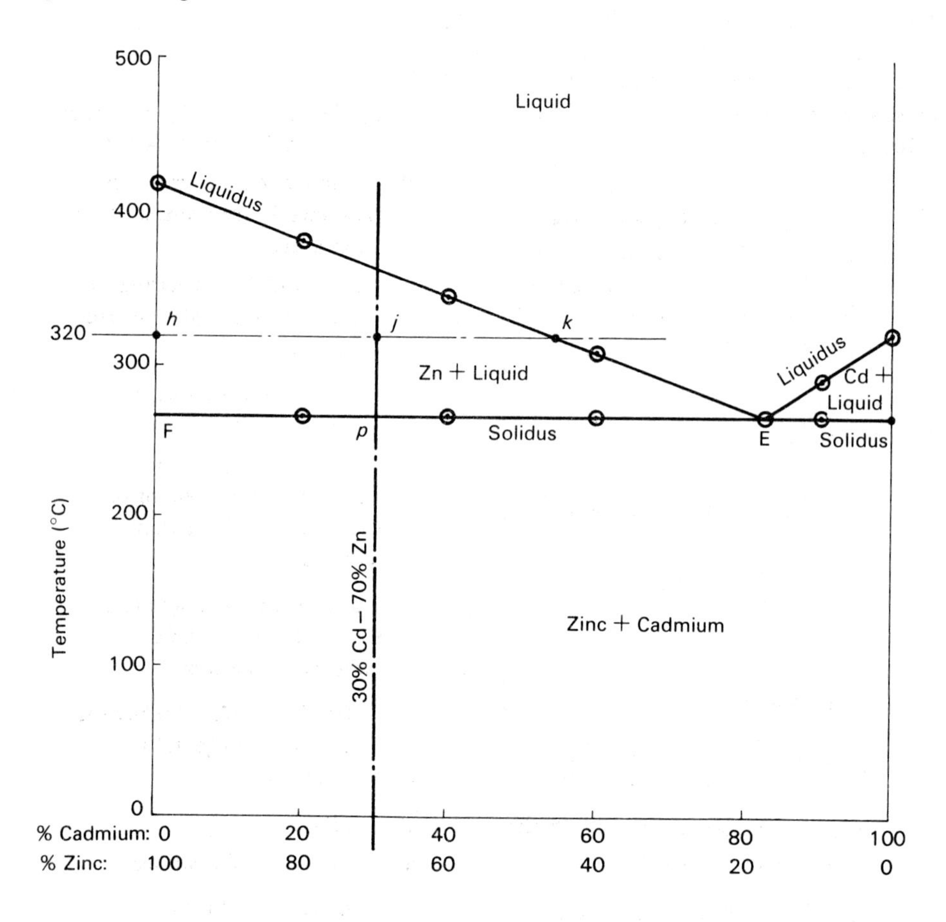

Fig. S.2

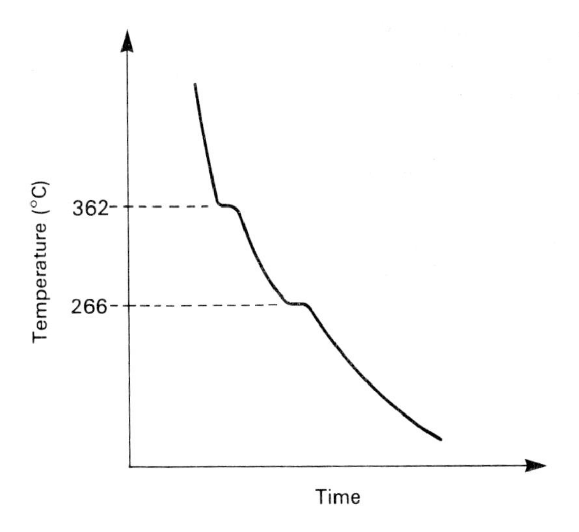

Fig. S.3

From the equilibrium diagram, the cooling curve for the 30% Cd, 70% Zn alloy would be of the form shown, with a first arrest at 362°C and a second arrest at 266°C.

(a) At 320°C, there will be dendrites of pure zinc in a liquid whose composition is given by the point k; i.e. the composition of the liquid will be 53.5% Zn, 46.5% Cd.

(b) At 320°C,

$$\text{Ratio}\left(\frac{\text{Solid zinc}}{\text{Liquid of composition } k}\right)$$

$$= \frac{jk}{hj} = \frac{23.5}{30} = \frac{0.78}{1}$$

(c) At the solidus,

$$\text{Ratio}\left(\frac{\text{Solid zinc}}{\text{Eutectic}}\right) = \frac{\text{E}p}{\text{F}p} = \frac{53}{30}$$

Therefore the proportion of eutectic in the final grain structure

$$= \frac{30}{30+53} = \frac{30}{83} = 0.36 \text{ or } 36\%$$

EXERCISES

Exercise 1

1) (b) By the number of electrons they have in their normal state.

(c) The number of electrons and vacancies for electrons in their outer shells.

(d) (ii) 7. (iii) Two electrons in innermost shell; eight electrons in next shell.

2) (a) Long chains of repeating patterns of atoms.

(b) Positively charged atoms permeated by a cloud of outer-shell electrons in random motion.

3) (a) The atoms are arranged in a repeated, simple, three-dimensional pattern.

4) (a) (i) See Fig. 1.12. (ii) See Fig. 1.6. (iii) See Fig. 1.10.

(b) See Fig. 1.8.

5) (a) See Fig. 1.13.

(b) Allotropic modifications.

6) (a) (i) f.c.c. (ii) b.c.c.
(ii) c.p.h.

7) (a) See Fig. 1.14.
   (b) See Fig. 1.15.

Exercise 2

8) (a) See Figs. 1.6, 1.10 and 1.12.
   (b) 14, 17 and 9.
   (c) See p. 10.

9) (b) (i) 522 °C (ii) 342 °C
   (iii) Liquid contains 24% antimony, 76% bismuth; Solid contains 73% antimony, 27% bismuth.
   (iv) Solid : liquid = 1.13 :1.

10) (b) (i) Solid solution: 45% A, 55% B; Liquid solution: 20% A, 80% B.
   (ii) Ratio $\left(\frac{\text{Solid solution}}{\text{Liquid solution}}\right) = \frac{1}{4}$
   (c) (i) As Fig. 2.8 (c): equi-axed grains of solid solution of 25% A, 75% B.
   (ii) As Fig. 2.10(a): layered dendrites: central layers of composition approximately 50% A, outer layers decreasingly rich in A but exceeding 25% A. Inter-dendritic grains richer in B than 75% B.
   (d) See Fig. 2.5.

11) The 70% A alloy has 1st arrest at 1070 °C; 2nd at 700 °C.
   (a) At 900 °C, the constituents are dendrites of solid pure A in a liquid consisting of 54% A, 46% B.
   (b) Solid : liquid = 0.53 : 1;
   (c) 46%;
   (d) 35% A, 65% B.

12) (a) 207 °C.
   (b) Dendrites of solid pure bismuth in a liquid consisting of 29% cadmium, 71% bismuth.
   (c) Solid 31%, liquid 69%.
   (d) 50%.

13) The shape of the diagram will be similar to that of Fig. 2.19.

15) (a) Shape of diagram similar to Fig. 2.19.
   (b) 40% A, 60% B.
   (c) 10%.
   (d) 20%.

Exercise 3

3) (b) 810 °C and 700 °C.
   (c) Similar to Fig. 3.4(b).

4) (a) $\frac{\text{Ferrite}}{\text{Pearlite}} = \frac{3.15}{1}$
   (b) $\frac{\text{Ferrite}}{\text{Pearlite}} = \frac{0.38}{1}$

For the remaining exercises there are no calculations, and hence no numerical 'answers'. However, where the answer to a question can be briefly indicated by reference to a page or diagram number, this has been done, as follows:

7) Similar to upper diagram of Fig. 3.6.

9) (a) Similar to Fig. 3.4(c).
   (b) Similar to Fig. 3.4(d).

10) (a) About 1.4% C.
   (b) About 0.1% C.
   (c) About 0.8% C.

Exercise 4

3) See Fig. 4.3.

5) (a) See Fig. 4.2(b).
   (b) See Fig. 4.2(c).

8) See Fig. 4.8.

11) See pp. 74–5

12) See Fig. 4.11.

14) 0.6–0.8% C; heat to 840 °C; oil quench; temper.

Exercise 5

2) See p. 83.

3) Silicon, sulphur, phosphorus, manganese.

6) (a) See Fig. 5.2(a).
(b) See Fig. 5.2(b).

7) See Fig. 5.2(d).

9) See Table 5.2, p. 94.

10) See p. 90.

Exercise 6

1) See p. 96.

2) See p. 96.

9) See Fig. 6.7.

10) See Fig. 6.6.

23), 24) See pp. 120–2.

25) See p. 121.

Exercise 7

2) See Figs. 7.6 and 7.7.

4) See Fig. 7.12.

5) See Figs. 7.9 and 7.10.

SOLUTIONS TO CROSSWORDS

1) *Across:* 1 RIP; 3 BCC.
Down: 1 RIB; 2 PVC.

2) *Across:* 1 ETCH; 4 PA; 5 SR; 6 SAND.
Down: 1 EPNS; 2 TA; 3 HARD; 5 SN.

3) *Across:* 1 BRINE; 4 ATOMS; 5 SISAL.
Down: 1 BRASS; 2 IRONS; 3 EASEL.

4) *Across:* 1 SOLVUS; 4 ANNEAL; 5 DARKER.
Down: 1 SOAKED; 2 LINEAR; 3 SOLDER.

5) *Across:* 1 ELEMENT; 4 CARBON; 7 SLIDE; 8 TENDERS.
Down: 1 ETCHANT; 2 EBONITE; 3 TI; 5 RESIN; 6 JETS.

6) *Across:* 2 CORES; 5 ZERO; 6 GRAPHITE; 7 LIQUIDUS; 9 DENDRITE.
Down: 1 PEARLITE; 2 COMPOUND 3 REMIND; 4 STRESS; 8 ULT.

7) *Across:* 1 ALPHA-IRON; 4 CAMPUS; 5 PC; 7 FIR; 8 DUCTILITY; 9 IN; 10 DIAL; 12 EUREKA; 13 MONOMER.
Down: 2 PUMICE; 3 ALUMINIUM; 4 CADMIUM; 5 PIT; 6 CRYSTAL; 7 FILLER; 10 DEO; 11 ARE.

8) *Across:* 5 FERRITE; 6 HE; 7 MICROSCOPE; 9 DELTA-IRON; 10 RED; 11 JOMINY TEST.
Down: 1 BRANCHED; 2 DISTORTION; 3 FES; 4 PHOSPHORUS; 5 FLAME; 8 CLIENT; 9 DEMO.

9) *Across:* 1 BCC; 3 COPPER; 7 CRYSTALLINE; 9 PRECIPITATE; 10 MILD STEEL; 11 SI; 12 MG; 13 BRINE.
Down: 2 CARBURISING; 4 POLYSTYRENE; 5 EIN; 6 STABILISER; 8 EUTECTOID.

10) *Across:* 4 SEAL; 7 UNIT COST; 9 FAR; 10 CASE REFINING; 11 AR; 12 SHORTERISING; 16 PRE-GRINDER; 18 REDUCING; 19 URN; 21 MO; 22 DIAGRAM.
Down: 1 ANNA; 2 PTFE; 3 CORE REFINING; 5 EQUILIBRIUM; 6 LONG; 8 CARAT; 10 COS; 13 HARDEN; 14 RIG; 15 GRAIN; 17 DOG; 20 ROW.

INDEX